Bioresorbable Polymers

Also of interest

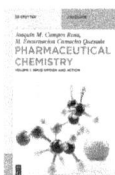

Pharmaceutical Chemistry.
Volume 1: Drug Design and Action
Campos Rosa, Camacho Quesada, 2017
ISBN 978-3-11-052836-7, e-ISBN 978-3-11-052848-0

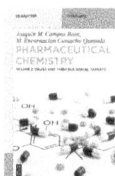

Pharmaceutical Chemistry.
Volume 2: Drugs and Their Biological Targets
Campos Rosa, Camacho Quesada, 2017
ISBN 978-3-11-052851-0, e-ISBN 978-3-11-052852-7

Magneto-Active Polymers.
Fabrication, characterisation, modelling and simulation
at the micro- and macro-scale
Pelteret, Steinmann, 2019
ISBN 978-3-11-041951-1, e-ISBN 978-3-11-041857-6

Bioelectrochemistry.
Design and Applications of Biomaterials
Cosnier (Ed.), 2019
ISBN 978-3-11-056898-1, e-ISBN 978-3-11-057052-6

e-Polymers.
Editor-in-Chief: Seema Agarwal
ISSN 2197-4586
e-ISSN 1618-7229

Bioresorbable Polymers

Biomedical Applications

Edited by
Declan M. Devine

2nd Edition

DE GRUYTER

Editor
Dr. Declan M. Devine
Materials Research Institute
Athlone Institute of Technology
Dublin Rd
Athlone,
Co. Westmeath
Ireland

ISBN 978-3-11-064056-4
e-ISBN (PDF) 978-3-11-064057-1
e-ISBN (EPUB) 978-3-11-064066-3

Library of Congress Control Number: 2019931987

Bibliographic information published by the Deutsche Nationalbibliothek
The Deutsche Nationalbibliothek lists this publication in the Deutsche Nationalbibliografie;
detailed bibliographic data are available on the Internet at http://dnb.dnb.de.

© 2019 Walter de Gruyter GmbH, Berlin/Boston
Typesetting: Integra Software Services Pvt. Ltd.
Printing and binding: CPI books GmbH, Leck
Cover image: HYBRID MEDICAL ANIMATION / SCIENCE PHOTO LIBRARY

www.degruyter.com

Preface

Maurice Dalton, Martin Forde, Ian Major and Declan Devine

Bioresorbable or biodegradable polymers are commonly used in various biomedical applications. Biodegradable polymers such as polyglycolic acid were first introduced into biomedical science in the 1960s, followed by polylactic acid (PLA) and polydioxanone in 1981 as suturing monofilament threads. The application of bioresorbable polymers in the biomedical sector has been widely exploited since then by immobilising suturing thread with analgesic or antibacterial drugs, and the development of bioresorbable vascular scaffolds, wound-healing and intravenous drug-delivery devices. Furthermore, biodegradable polymers have been investigated as a replacement for metallic orthopaedic devices due to their precise control of material composition and microstructure. Other bioresorbable polymers such as polyethylene oxide dissolve in physiological fluid without breaking down of their macromolecular chains. These polymers are eliminated from the body *via* dissolution, assimilation and excretion through metabolic pathways. The hydrolysing process breaks down the polymer into smaller units and its degradation products are excreted by means of the citric acid cycle or by direct renal excretion [1]. Vert and co-workers [2] stated that the metabolites of bioresorbable materials are eliminated from the body with no residual side effects *via* assimilation of monomer and oligomers units. The backbone remains intact due to the absence of phagocytotic cells [2, 3]. The degradation process between natural and synthetic resorbable polymers can be hydrolytic or enzymatic. Essentially, most synthetic polymers are degraded by hydrolytic degradation, whereas natural polymers degrade *via* enzymatic degradation. Kim and co-workers indicated that the degradation of synthetic resorbable polymers is easier to predict because water levels are constant in physiological conditions. However, natural polymers rely on enzymatic degradation, making degradation unpredictable because enzyme levels differ from tissue to tissue and from patient to patient [4].

Processing of bioresorbable implants can be achieved *via* conventional polymer processing methods such as extrusion, injection and compressing moulding, solvent spinning or casting. However, special consideration must be given when processing these materials because heat can cause a reduction in the molecular weight (MW) due to the hydrolysing of bonds. For example, Maspoch and co-workers [5] characterised PLA extensively and concluded that processing PLA *via* thermoplastic processes leads to decreases in MW [5, 6]. In addition, overheating can depolymerise the polymer and, as a result, monomers can have a plasticising effect on the polymer [7]. Recently, alternative approaches utilising rapid prototyping and micro-/nanofabrication processes have been employed. These methods allow for improvement in the control of the microstructure of biomaterial scaffolds [8].

https://doi.org/10.1515/9783110640571-201

The current work aims to address these issues and to highlight recent advances in the biomedical field that have being enabled by the use of biodegradable polymers.

References

1. A.S. Hoffman, Advanced *Drug Delivery Reviews*, 2012, **64**, 18.
2. C.X. Lam, M.M. Savalani, S.H. Teoh and D.W. Hutmacher, *Biomedical Materials*, 2008, **3**, 34108.
3. M. Vert, S.M. Li, G. Spenlehauer and P. Guerin, *Journal of Materials Science:* Materials *in Medicine*, 1992, **3**, 432.
4. S.H. Kim and Y. Jung in Biotextiles as Medical Implants, Eds., M.W. King, B.S. Gupta and R. Guidoin, Woodhead Publishing Ltd, Cambridge, UK, 2013.
5. F. Carrasco, P. Pags, J. Gmez-Prez, O.O. Santana and M.L. Maspoch, *Polymer Degradation and Stability*, 2010, **95**, 116.
6. S. Farah, D.G. Anderson and R. Langer, *Advanced Drug Delivery Reviews*, 2016, **107**, 367.
7. K.J. Burg, S. Porter and J.F. Kellam, *Biomaterials*, 2000, **21**, 2347.
8. G. Narayanan, V.N. Vernekar, E.L. Kuyinu and C.T. Laurencin, *Advanced Drug Delivery Reviews*, 2016, **107**, 247.

Contents

Gavin Burke, Elaine Kenny, Maurice Dalton, Declan M. Devine, Eilish Hoctor,
Ian Major and Luke Geever

Yuanyuan Chen, Marcelo Jorge Cavalcanti De Sá, Maurice Dalton
and Declan M. Devine

Gabriel Goetten de Lima, Shane Halligan, Luke Geever, Maurice Dalton,
Chris McConville and Michael J.D. Nugent

Martin Forde and Ian Major

Ian Major, Elaine Kenny, Andrew Healy, Luke Geever, Declan M. Devine
and John Lyons

Emily Crowley, Maurice Dalton and Gavin Burke

Contributors

Gavin Burke
Athlone Institute of Technology,
Materials Research Institute,
Dublin Road, Athlone,
County Westmeath,
N37 HD68, Ireland

Marcelo Jorge Cavalcanti De Sá
Athlone Institute of Technology,
Materials Research Institute,
Dublin Road, Athlone,
County Westmeath,
N37 HD68, Ireland

Yuanyuan Chen
Athlone Institute of Technology,
Materials Research Institute,
Dublin Road, Athlone,
County Westmeath,
N37 HD68, Ireland

Emily Crowley
Athlone Institute of Technology,
Materials Research Institute,
Dublin Road, Athlone,
County Westmeath,
N37 HD68, Ireland

Maurice Dalton
Athlone Institute of Technology,
Materials Research Institute,
Dublin Road, Athlone,
County Westmeath,
N37 HD68, Ireland

Declan M. Devine
Athlone Institute of Technology,
Materials Research Institute,
Dublin Road, Athlone,
County Westmeath,
N37 HD68, Ireland

Martin Forde
Athlone Institute of Technology,
Materials Research Institute,

Dublin Road, Athlone,
County Westmeath,
N37 HD68, Ireland 'Gabriel Goetten de Lima
and address'

Gabriel Goetten de Lima
Athlone Institute of Technology,
Materials Research Institute,
Dublin Road, Athlone,
County Westmeath,
N37 HD68, Ireland

Luke Geever
Athlone Institute of Technology,
Materials Research Institute,
Dublin Road, Athlone,
County Westmeath,
N37 HD68, Ireland

Shane Halligan
Athlone Institute of Technology,
Materials Research Institute,
Dublin Road, Athlone,
County Westmeath,
N37 HD68, Ireland

Andrew Healy
Athlone Institute of Technology,
Materials Research Institute,
Dublin Road, Athlone,
County Westmeath,
N37 HD68, Ireland

Eilish Hoctor
Athlone Institute of Technology,
Materials Research Institute,
Dublin Road, Athlone,
County Westmeath,
N37 HD68, Ireland

Elaine Kenny
Athlone Institute of Technology,
Materials Research Institute,
Dublin Road, Athlone,

https://doi.org/10.1515/9783110640571-202

County Westmeath,
N37 HD68, Ireland

John Lyons
Athlone Institute of Technology,
Materials Research Institute,
Dublin Road, Athlone,
County Westmeath,
N37 HD68, Ireland

Ian Major
Athlone Institute of Technology,
Materials Research Institute,
Dublin Road, Athlone,
County Westmeath,
N37 HD68, Ireland

Chris McConville
School of Pharmacy,
Institute of Clinical Sciences,
Sir Robert Aitken Institute for Medical
Research,
University of Birmingham, Edgbaston,
Birmingham, B15 2TT, UK

Michael J.D. Nugent
Athlone Institute of Technology,
Materials Research Institute,
Dublin Road, Athlone,
County Westmeath,
N37 HD68, Ireland

Gavin Burke, Elaine Kenny, Maurice Dalton, Declan M. Devine,
Eilish Hoctor, Ian Major and Luke Geever

1 Biodegradation and biodegradable polymers

1.1 Introduction

Polymers are organic materials consisting of large macromolecules composed of many repeating units called 'mers'. These long molecules are covalently bonded chains of atoms. Unless they are crosslinked, the macromolecules interact with one another by weak secondary bonds (hydrogen bonds and van der Waals forces) and by entanglement. The mechanical and thermal behaviour of polymers is influenced by several factors, including the composition of the backbone and side groups, the structure of the chains, and the molecular weight (MW) of the molecules.

'Biodegradable polymers' can be defined as polymers that are degradable *in vivo*, either enzymatically or non-enzymatically, to produce biocompatible or non-toxic by-products. These polymers can be metabolised and excreted *via* normal physiological pathways [1]. In recent years, emphasis in biomaterials engineering has moved from materials that remain stable in the biological environment to materials that can degrade in the human body. Biodegradable materials are designed to degrade gradually and be replaced eventually by newly formed tissue in the body. Biodegradable implants have the advantage of allowing the new tissue, as it grows naturally, to take over load-bearing or other functions without any of the potential chronic problems associated with bio-stable implants [2].

'Biodegradation' is used to describe the process of a material being broken down by nature. However, in the case of medical-purpose biomaterials, biodegradation focuses on the biological processes within the body that cause a gradual breakdown of the material. Biomaterials degradation is a very important aspect to consider if they are used for medical purposes because their ability to function for a certain application depends on the length of time needed to keep them in the body [3, 4]

The simple desire of a physician is to have a device that can be used as an implant and does not necessitate a second surgical procedure for removal. In addition to not requiring a second surgical procedure, the biodegradation may offer other advantages. For example, a fractured bone stabilised with a rigid, non-biodegradable stainless-steel implant has a tendency to weaken the bone and re-fracture (i.e., stress shielding). The bone does not carry a sufficient load during the healing process because the load is carried by the rigid stainless steel. However, an implant prepared from biodegradable polymer(s) can be engineered to degrade at a rate that will slowly transfer load to the healing bone [5, 6]. Also, the degradation and resorption kinetics must be controlled in such a way that the bioresorbable scaffold retains its

https://doi.org/10.1515/9783110640571-001

physical properties and thereafter begins to lose its mechanical properties, subsequently being metabolised by the body without a reaction to a foreign body. Apart from the anatomy and physiology of the host, the type of tissue to be engineered has a profound influence on the degree of remodelling. For example, in cancellous bone, the remodelling takes 3–6 months, whereas cortical bone takes ≈6–12 months to remodel. Whether the biodegradable construct will be part of a load-bearing or non-load-bearing site will also significantly influence the needs for mechanical stability of the biodegradable construct because mechanical loading can affect the degradation behaviour directly. A precise timeline for the biodegradation of these materials has been reported inconsistently in the literature. For example, Hutmacher [7] stated that a scaffold must retain its strength for ≥6 months and thereafter gradually reduce over 12–18 months. This finding is similar to that of a recently published study by Gómez-Barrena and co-workers [8], who stated that inadequate formation of bone 6 months after scaffold implantation can be declared a 'non-union'. Li and co-workers [9] and Kattimani and co-workers [10] reported bone regrowth to have occurred after 3 months, with the latter study stating that grafts healed fully within 8 weeks of implantation, followed by steady bone density after 3 months. However, Mäkinen and co-workers [11] stated that, based on the literature and experience with orthopaedic implants, a scaffold should be resorbed between 1 and 5 years.

The prevailing mechanism for polymer degradation is chemical hydrolysis of the hydrolytically unstable backbone. In semicrystalline polymers, chemical hydrolysis can be divided into two phases. Initially, water molecules infiltrate the polymer structure and selectively target the amorphous part of the polymer. This process results in the breakdown of long polymer chains into shorter, more hydrophilic polymer chains. Once this fragmentation occurs, the polymer suffers a reduction in MW and mechanical strength. This process leads to the second phase, in which enzymatic breakdown of the polymer fragments results in rapid breakdown of the polymer [12–14].

Secondary to chemical hydrolysis is bulk erosion. This process occurs if the rate of water penetration occurs faster than the polymer can be converted into water-soluble molecules. In this scenario, there is an initial increase in the rate of surface degradation whereas the interior environment of the polymer becomes increasingly acidic with the increasing concentration of polymer-degradation products, which cannot diffuse out of the polymer. This increase in acidity then creates a self-catalysed degradation of the polymer. This form of polymer degradation has been utilised to control polymer-degradation rates [15, 16]. If the opposite is true (i.e., the rate of water penetration is slower than the rate of polymer conversion to water-soluble molecules), then the third mode of polymer degradation, surface erosion, occurs [17]. This polymer variant results in a slow decrease in polymer size with consistent mechanical strength throughout its structure. Common examples of this type of variant are polyanhydrides and poly-orthoesters, wherein the polymers are hydrophobic but the chemical bonds are highly susceptible to hydrolysis.

'Controlled biodegradation' is a critical factor in developing tissue scaffolds that can be gradually reabsorbed by and excreted from the body. In general, biodegradability depends on the following factors [18]:
1. The chemical stability of the polymer backbone
2. The presence of catalysts
3. Additives, impurities or plasticisers
4. The geometry of the device
5. The location of the device
6. Hydrophobicity of the monomer
7. The morphology of the polymer
8. The initial MW
9. Fabrication processes
10. Properties of the scaffold, such as porosity and pore diameter

Furthermore, certain factors can accelerate the rates at which polymer degradation takes place. These factors must be accounted for when seeking to 'tailor' a controllably degradable biomedical product:
1. The hydrophilicity of a monomer
2. The hydrophilicity of acidic end groups
3. The number of reactive groups present in the polymer backbone
4. Degree of polymer crystallinity
5. Size of the polymer device

Another key factor in the degradation rates of implants is the *in vivo* environment. If a device is located in a poorly vascularised region and produces sufficiently acidic by-products, then a localised decrease in pH can occur. This localised decrease in pH can then further catalyse the degradation of the implant, causing a sharp increase in localised acidity. This localised decrease in pH has been identified as a cause of an adverse tissue response [19, 20]. Another consideration for biomedical devices is localised stress because implants under stress degrade faster, possibly due to the formation of 'microcracks' that create an increased surface area in the polymer [19, 21, 22].

In this regard, the MW has a major role in the degradation behaviour. A large molecular weight distribution (MWD) would indicate relatively large numbers of carboxylic end groups which can facilitate the autocatalytic degradation of the polymer chains. A large or wide MWD, thus, would be expected to accelerate the rate of degradation whereas a small or narrow MWD would have fewer carboxylic acid end groups available for autocatalysis. Other factors, such as the MW, MWD and sterilisation, may also alter the degradation rate of the biodegradable polyesters in microspheres [23].

The most important biomedical applications of biodegradable polymers are controlled drug-delivery systems [24] as well as implants and devices for bone and dental repairs [25]. The internal fixation devices used in orthopaedic surgery serve no purpose as soon as they have fulfilled their mission of securing the healing and union of the tissues concerned. Biodegradation has been accomplished by synthesising polymers that have hydrolytically unstable linkages in the backbone. Ideally, biodegradable polymers would:

1. Be metabolised in the body and eliminated by normal physiological pathways;
2. Be fabricated easily into their final forms;
3. Be degraded into non-toxic substances that are non-mutagenic and non-cytotoxic;
4. Cause no initiation of inflammatory processes after application, injection or insertion;
5. Be easily sterilised; and
6. Have an acceptable shelf-life [16].

An important criterion to be considered for a polymer to be suitable as a biodegradable polymer is the end product after degradation. The end products of aerobic degradation from biodegradable polymers should be carbon dioxide (CO_2), water and/or minerals [26].

Polymer materials differ in their MW, polydispersity, crystallinity, and thermal transition, and different degradation rates that would strongly affect polymer scaffold properties. For example, polymer hydrophobicity and percent crystallinity have an effect on the cellular phenotype, and deflection in the surface charges affect the cellular spreading [15]. Synthesised biodegradable polymers have received increasing interest owing to the difficulty in obtaining reproducibility when using natural polymers. Biodegradable synthetic polymer materials such as polyglycolic acid (PGA), polylactic acid (PLA), and their copolymers, poly(p-dioxanone), and copolymers of trimethylene carbonate and glycolide have been used in several clinical applications such as resorbable sutures, drug-delivery systems and orthopaedic fixation devices (e.g., pins, rods and screws) [27]. In general, bioresorbable materials can be categorised into two distinctive groups: 'natural' and 'synthetic'.

1.2 Natural biodegradable polymers

Natural polymers are derived from animal tissue or plant fibres. These materials possess unique characteristics in which they do not inflict an immunological response within the host and toxic degradation is not produced. They can closely mimic the natural extracellular matrix of tissues, have low immunogenic potential, can interact with host tissues, and have chemical versatility. Natural biodegradable polymer materials are derived from proteins such as collagen, gelatin and albumin and polysaccharides such as cellulose, hyaluronate, chitin and alginate (Table 1.1). Natural polymers such as collagen, chitin, hyaluronic acid and silk have all been assessed for wound dressing and use as temporary barriers. In some cases, there is also an unlimited supply of the source. Alginate and chitosan (CS) are two natural polysaccharides that do not exist

Table 1.1: Main families of natural biodegradable polymers.

Polysaccharides		Starch
		Pullulan
		Cellulose
		Pectin
		Xanthan
		Lignin
		Chitin
		CS
		Dextran
		Gellan
Polyesters	Microbial origin	PHA
		PHB
	Biotechnological origin	PLA
Proteins		Wheat and corn gluten
		Polypeptides of aspartic acid and lysine
		Casein
		Soy
		Collagen and gelatin

PHA: Polyhydroxyalkanoates
PHB: Polyhydroxysbutyrate

within the human body but have been investigated for tissue-engineering applications because they are structurally similar to the glycosaminoglycans found in the natural extracellular matrix of tissues (i.e., skin, bone, and blood vessels). Alginate originates from seaweed and is attractive because of its low toxicity, water solubility and its simple gelation chemistry with calcium ions. Alginate hydrogels have been investigated for use as scaffolds for cartilage and liver regeneration [28], as well as for wound dressings. CS is a derivative of naturally occurring chitin, which is found in the exoskeletons of crustaceans. It has low toxicity and is biocompatible. CS scaffolds have been investigated for skin and bone-tissue engineering [29]. CS exhibits a wide array of unique characteristics: good tensile strength, a minimal inflammatory response and cellular targeting sites. However, the production cost associated with preparing natural material for medical applications is high [30].

1.3 Synthetic biodegradable polymers

The first synthetic biomaterial was introduced commercially as a biodegradable suture (Dexon®) in the late-1960s. Since then, expanding use of synthetic bioresorbable polymers has led to experimental and clinical applications in orthopaedics, pharmaceuticals, drug delivery and, more recently, scaffolds for tissue engineering [31]. The mechanical, physical and biological properties of synthetic polymers can be tailored to give a wide range of controllable properties that are more predictable than materials obtained from natural sources. Synthetic materials can be customised by adjusting the ratios of the monomer units (basic building blocks of the final polymer) and by the incorporation of specific groups or molecules (e.g., Arginyl-glycyl-aspartic acid peptide that cells can recognise). The degradation rate and products can be controlled by appropriate selection of the segments to form breakdown products that can be metabolised into harmless by-products or can be excreted *via* the renal filtration system. Among the many biodegradable synthetic polymers used for biomedical applications, one particular group of synthetic bioresorbable polymer that is investigated widely is aliphatic polyesters comprising PGA, PLA, polydioxanone (PDO) and polycaprolactone (PCL). These aliphatic polyesters are utilised due to their improved mechanical properties, including degradability, processability and, most importantly, they are biocompatible [32, 33]. Polyesters such as PGA, poly(L-lactic acid) (PLLA) and poly(D-lactic acid) (PDLA) are commonly used in biomedical applications because their degradation products are physiological metabolites due to non-enzymatic hydrolysis (Table 1.2). The degradation is dictated by the monomer structure, MW, crystallinity and substituting groups [34–36]. These polymers degrade by hydrolytic mechanisms and are commonly used because their degradation products can be removed from the body as CO_2 and water. These materials are discussed later in this book.

Table 1.2: Main families of synthetic biodegradable polymers.

Polyesters	Aliphatic	PCL
		Phosphate-buffered saline
		PLA
		PGA
	Aromatic	Polybutylene terephthalate
Polyvinyl alcohol		

Traditionally, polyurethanes (PU) were used in the biomedical field as blood-contact materials for cardiovascular devices, and were intended to be used as non-degradable coatings. More recently they have been designed to be biodegradable by being combined with degradable polymers such as PLA for soft tissue-engineering applications. However, there is a lowering of the pH in the localised region, resulting in an inflammatory response when they degrade. PCL has a very similar structure to PLA and PGA and is also degraded *via* hydrolytic mechanisms under physiological conditions. In addition, it is degraded enzymatically and the resulting low-MW fragments are reportedly taken up by macrophages and degraded intracellularly. PCL is predominantly used for drug-delivery devices because it has a slower degradation rate than PGA and PLA. However, more recently, it is increasingly finding applications in tissue engineering.

Polyethylene glycol (PEG) is a biocompatible, non-toxic, water-soluble polymer that is a liquid at cold temperatures and elastic gel at 37 °C [37]. PEG-based copolymers have been used as injectable scaffolds for bone as well as for drug-delivery applications [29]. Polyethylene oxide (PEO) has been assessed extensively as a bioresorbable carrier for oral drug-delivery devices, and it has been reported that various fillers can be used to tailor the release of the active pharmaceutical ingredients (API). Lyons and co-workers demonstrated that the use of agar as a filler material in hot-melt extrusion dosage forms leads to altered release profiles of diclofenac sodium as a result of increasing viscosity in the PEO matrix [38]. Similarly, the inclusion of nanoclays not only improved the mechanical properties of the polymeric matrix, but also permitted slow release of API in comparison with the matrix alone [39]. Furthermore, supercritical fluid technology utilising CO_2 during hot-melt extrusion has been shown to act as a plasticiser in the production of monolithic dosage forms and, due to increased porosity of the scaffold, drug release is increased [40].

Copolymers of PEG and PLA have been created whereby the degradation rate and hydrophilicity can be controlled by adjusting the ratio of the hydrophilic PEG to hydrophobic PLA blocks [41]. Synthetic biodegradable polymeric materials offer more advantages over natural materials in that they can be synthesised to give various properties, such as reproducibility and no immunogenicity concerns. Also, they are a reliable source of raw materials because the polymer materials, with their fundamental building block units, have simple and well-known structures and properties (Table 1.3). Synthetic materials are often selected as tissue-engineering materials to produce scaffolds because their polymeric molecules are available commercially and there is no need for special processing before use [42].

1.4 Polylactic acid

PLA (Figure 1.1) is the most studied aliphatic polyester and demonstrates potential in biomedical applications such as resorbable surgical sutures, meshes or in drug delivery [44]. PLA is a semicrystalline thermoplastic aliphatic polyester that has fragile characteristics, including a high elastic modulus and low elongation at break and, as a result, is preferred for bone engineering. This thermoplastic polymer possesses unique characteristics, such as high strength, biocompatibility, and monomeric units within the polymer that can be synthesised *via* non-toxic renewable feedstock as well as being derived naturally [43, 45]. The initial step in the degradation of PLA is hydrolysing the implantable PLA into intermediate α-hydroxy acid, which is broken down metabolically *via* the tricarboxylic acid cycle and excreted. Fukushima

Table 1.3: Bioresorbable polyesters (including structures and terminal properties) used in tissue engineering.

Precursor	Structure	Polymer	Structure	T_g (°C)	T_m (°C)	Degradation time	Degradation products	Structure of degradation product	Ref.
Lactides		PLA		45–60	150–162	Several years	Lactic acid		[32, 43]
Glycolide		PGA		35–45	220–230	6–12 months	Glycolic acid		[32, 43]
ε-caprolactone		PCL		−60	59–64	Several years	Caproic acid		[32, 43]

T_g: Glass transition temperature
T_m: Melting temperature

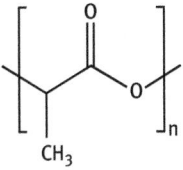

Figure 1.1: Chemical structure of PLA.

and co-workers studied the degradation mechanism of PLA and concluded that incorporation of increased amounts of nanoclays increased PLA degradation *via* bacterial degradation [46]. In addition, Chen and co-workers demonstrated that increased amounts of halloysite can alter the thermal stability of PLA, which was associated with the presence of voids between the polymer and matrix [47].

PLA degradation is non-toxic and does not have toxic or carcinogenic effects in local tissue. PLA possesses better thermal properties in comparison with PCL, and as a result can be processed much easier *via* injection moulding, extrusion, blow moulding, thermoforming and fibre spinning [43]. However, PLA is hydrophobic and degrades slowly, which is an unfavourable characteristic in biomedical applications due to the possibility of initiating an inflammatory response.

1.5 Polycaprolactone

PCL (Figure 1.2) is a hydrophobic, semicrystalline polyester synthesised most commonly *via* splitting of a ε-caprolactone ring or through the polycondensation of 6-hydroxyhexanoic acid. Although these are the most common methods of synthesis, the number of methods available for the synthesis of PCL is extensive, and covers a large variety of catalytic systems [48]. PCL has good solubility in organic solvents, a T_m of 55–60 °C and a transition glass temperature (T_g) between −65 and −54 °C [49, 50]. Furthermore, it has a tensile strength of 23 MPa and very good elongation before break of >700% [51].

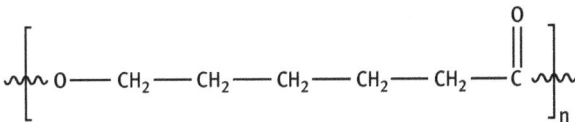

Figure 1.2: Chemical structure of PCL.

First developed in the early 1930s, PCL was one of the major polymers used for drug-delivery systems throughout the 1970s and 1980s due to its tailorable degradation kinetics and mechanical properties as well as ease of shaping during manufacture. For example, Darney and coworkers [52] utilised the long degradation rate of PCL (2–3 years) to produce a long-term contraceptive that is still in use today. Though initially viewed as highly desirable, PCL was subsequently overshadowed by polymers such as PGA and PLA due to their faster degradation rates and higher load-bearing properties. In more recent years there has been a resurgence in the popularity of PCL owing to its superior rheological and viscoelastic properties when compared with other resorbable polymers, which allow for greater flexibility in the creation of tissue-engineering scaffolds [53]. Furthermore, there has been investment in the development of micro and nanoscale drug-delivery vehicles. However, the PCL degradation rate has been a significant issue when trying to gain approval for pure PCL products by the US Food and Drug Administration (FDA) [49] even though it is known that PCL degradation can be enhanced enzymatically [54]. Interestingly, PCL is miscible with many other polymers, including polyvinyl chloride, poly(styrene–acrylonitrile), poly(acrylonitrile–butadiene–styrene), poly (bisphenol A) and other polycarbonates (PC), nitrocellulose and cellulose butyrate, as well as being mechanically compatible with others, including polyethylene, polypropylene, natural rubber, polyvinyl acetate and polyethylene–propylene rubber [48]. This, added to the limitations of the mechanical properties and degradation rates of pure PCL, has led to extensive research into many PCL copolymers and blends with other polymers, including PLLA [55, 56], PDLA [57, 58] and poly(lactic-co-glycolic acid) (PLGA) [59, 60].

One of the key properties of PCL along with its good elongation and tensile strength is its degradation rate. While PCL is considered bioresorbable, the process by which it is metabolised and subsequently excreted is extremely slow. MW plays an important part in determining the degradation rates of PCL. However, it is the degradation mechanism itself which results in the slow breakdown of PCL *in vivo*, which is thought to be associated with the lack of suitable enzymes for such breakdown in the body. PCL degradation can be rapid in the presence of certain enzymes [61]. The commonly held view of PCL degradation is based on a combination of three processes: bulk erosion, bulk degradation and autocatalysis. In bulk erosion, the PCL surface becomes hydrated *in vivo*, and then the surface of the polymer scaffold is bulk-eroded by hydrolytic cleavage of the polymer backbone, resulting in the slow thinning of the polymer in a predictable manner. In bulk degradation, water penetrates the surface of the polymer, resulting in random hydrolytic cleavage within the polymer and resulting in a controlled decrease in the MW of PCL [62]. Autocatalysis occurs when the products of PCL degradation cause an internal acidic gradient due to an increase in formation of carboxyl end groups, resulting in an increased rate of reaction within the PCL scaffold. As the polymer chains become small enough to dissipate rapidly through the outer layer of the PCL scaffold, the latter is hollowed out [53]. Then, the end-products of PCL degradation can be excreted if the rate of release of acidic by products is above tolerable levels, but it can result in an inflammatory response [63].

Although limited to long-term drug-delivery applications due to its slow degradation rates, PCL has been shown to have good mechanical properties for certain applications as well as good processability [64, 65]. This has allowed for the development of many types of PCL, such as electrospun scaffolds [66], porous scaffolds [56] and nanosphere- and microsphere-embedded scaffolds. The variety with which PCL scaffolds can be manufactured has led to research in tissue-engineering scaffolds for the regeneration of nerves [64, 67], skin [68], cartilage [66, 69], ligaments, bone [60] and other tissues. Furthermore, PCL can enhance the properties of other polymers, as has been highlighted by Rodrigues [70] with regard to polymers such as polytrimethylene carbonate (PTMC), collagen, hyaluronic acid and CS.

1.6 Polyglycolic acid

PGA (Figure 1.3) is a rigid thermoplastic and was one of the earliest polymers studied for biomedical applications. Similar to PLA, PGA is a poly(α-hydroxy acid), being one of the most widely studied bioresorbable polymers available. PGA possesses a T_m of 225–230 °C, T_g of 35–40 °C [71] as well as an excellent tensile strength of 12.8 GPa [72]. PGA can be synthesised by direct polycondensation of glycolic acid, ring opening of glycolide (cyclic dimer of glycolic acid), oligomerisation of glycolic acid, and enzyme-catalysed reactions [73]. Due to the simplicity of its structure, PGA is typically a highly crystalline polymer, with crystallinity in the region of 45 to 55%, resulting in PGA having very low solubility in solvents [74] with the exception of highly fluorinated organic solvents such as hexafluoroisopropanol [51]. Furthermore, the simplicity of its backbone has resulted in issues with the manipulation of PGA polymer chains for the purpose of altering degradation rates, mechanical properties or polymer morphology [73]. However, this has also allowed for the development of numerous copolymers and polymer grafts of PGA, including PGA–PLA (which has attracted considerable interest in and of itself due to its biocompatibility, tailorable degradation rate (by varying PLA/PGA ratio) and its FDA approval for clinical use [75]), PGA–PCL (which has been discussed above) and PGA-collagen.

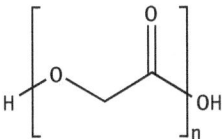

Figure 1.3: Chemical structure of PGA.

Known as a potential polymer for the synthesis of tough fibres since 1954, PGA is one of the earliest examples of a polymer developed for biomedical application. PGA first found use as a medical suture known as Dexon® in 1962 [76] due to its high tensile strength and in part to its inherent loss of >90% of its mechanical strength over 21 days caused by its rapid degradation [77]. PGA also found use as a bone pin under the commercial name Biofix from 1984 to 1996. This material was replaced with PLLA from 1996 onwards due to its better stability [78]. The stages of PGA degradation show similarities to those described for PLA (although PLA, as a slightly less hydrophilic polymer, takes slightly longer to degrade [79]) and PCL (which takes much longer [49]). This begins with the hydration of the polymer followed by the hydrolytic degradation of ester bonds that lower the mass of the polymer and allow for excretion [80] following the mechanism of bulk degradation. Furthermore, its degradation rate can also be increased by enzymatic degradation if needed using carboxypeptidase A, alpha-chymotrypsin, clostridiopeptidase A or ficin [81]. This is helped (at least in part) by the high degree of flexibility of polymer chains, which allows most aliphatic polyesters to fit into the active site of an enzyme, after which PGA degradation products are metabolised further to water and CO_2 or excreted through the kidneys [82]. PGA use in drug delivery has been limited by its rapid degradation, which increases the difficulty of carrying out controlled release of drugs and results in the rapid release of glycolic acid which, in turn, can cause tissue damage [51]. Outside of uses as a drug-delivery vehicle, the number of biomedical uses for PGA has expanded markedly. Non-woven PGA fabrics have garnered particular attention due to their excellent degradability, good mechanical strength and cell viability on matrices [49]. Tissues upon which PGA has been used to enhance regeneration include bone [83, 84], cartilage [85, 86], nerve [87, 88] and tendon [89, 90].

1.7 Polydioxanone

PDO is a colourless biodegradable polymer used in sutures [53]. PDO can be prepared from diethylene glycol [91] or through the ring opening of paradioxanone *via* a combination of heat and an organometallic catalyst such as zinc L-lactate [92]. PDO has been known in the industry for >60 years, with patents for its synthesis being filed as early as the 1950s [93] and further patents being published within the last 5 years [94]. Compared with other major biodegradable polymers, PDO is much less well known. This has been attributed by several authors to the fact that much of the information on PDO is protected by patents [92, 95]. PDO is unique in that it forms an ether-ester bond as opposed to an ester-ester bond, which is typically associated with biodegradable polymers. With a semicrystalline structure (≈55% crystallinity) PDO has a T_m of ≈110 °C [95] and a T_g of −10 °C to 0 °C [16]. This places its thermal properties between those found in PCL (T_m ≈ 60 °C) and PLA (T_m>175 °C [96]). PDO degrades fully over 6–12 months [16], allowing application as medical sutures. At present, a branch of Johnson & Johnson Company (Ethicon) markets the products PDO II suture and PDO plus antibacterial sutures. As some of the older patents reach their end of term, the number of PDO products available for therapeutic purposes will probably increase.

1.8 Polyhydroxyalkanoates

PHA are another group of polyesters, which due to their biodegradability and good biocompatibility, are attractive for tissue engineering. PHA are produced by microorganisms under unbalanced growth conditions. Thanks to their thermoprocessability and bioresorbable characteristics they have been used to develop devices such as sutures, suture fasteners, bone plates, cardiovascular patches, orthopaedic pins and nerve guides. The mechanical feature, biocompatibility and degradation rate of PHA can be adjusted by surface modification as well as combination with other polymers, enzymes or inorganic materials. Unfortunately, PHA production is limited due to their time-consuming production *via* microbes [31, 97].

1.9 Summary

The polymers discussed above are likely to be the most researched bioresorbable polymers available currently and they belong to the poly(α-ester) class of polymers. However, bioresorbable polymers such as polyanhydrides, polyacetals, PC and PU have been produced. Furthermore, there is considerable interest in the creation of enzymatically degradable polymers using polymers considered traditionally to be non-degradable, such as the polyethers. Development of more copolymers to overcome the limitations of each of the polymers mentioned above will be a major research area in the future.

Bibliography

1. J.A.Y.R. Joshi and R.P. Patel, *International Journal of Current Pharmaceutical Research*, 2012, **4**, 4, 74.
2. L.J. Chen and M. Wang, *Biomaterials*, 2002, **23**, 13, 2631.
3. M.A. Woodruff, C. Lange, J. Reichert, A. Berner, F. Chen,P. Fratzl, J.T. Schantz and D.W. Hutmacher, *Materials Today*, 2012, **15**, 10, 430.
4. E. Tamariz and A. Rios-Ramirez in *Biodegradation – Life of Science*,Veracruz, Mexico, 2013, p.3.
5. M.E. Gomes, A.S. Ribeiro, P.B. Malafaya, R.L. Reis and A.M. Cunha, *Biomaterials*, 2001, **22**, 9, 883.
6. K. Tuzlakoglu and R.L. Reis, *Tissue Engineering, Part B: Reviews*, 2009, **15**, 1, 17.
7. D.W. Hutmacher, *Biomaterials*, 2000, **21**, 24, 2529.

8. E. Gómez-Barrena, P. Rosset, D. Lozano, J. Stanovici, C. Ermthaller and F. Gerbhard, *Bone*, 2015, **70**, 93.
9. X. Li, N. Dunne, X. Li and K.E. Aifantis, *BioMed Research International*, 2014, **2014**, 2.
10. V.S. Kattimani, P.S. Chakravarthi, N.R. Kanumuru, V.V. Subbarao, A. Sidharthan, T.S.S. Kumar and L.K. Prasad, *Journal of International Oral Health*, 2014, **6**, 3, 15.
11. T.J. Mäkinen, M. Veiranto, P. Lankinen, N. Moritz, J. Jalava, P. Törmälä and H.T. Aro, *Journal of Antimicrobial Chemotherapy*, 2005, **56**, 6, 1063.
12. D. Ferguson, W.L. Davis, M.R. Urist, W.C. Hurt and P.E. Allen, *Clinical Orthapoedics and Related Research*, 1987, **219**, 251.
13. M.I. Sabir, X. Xu and L. Li, *Journal of Materials Science*, 2009, **44**, 21, 5713.
14. J. Henkel, M.A. Woodruff, D.R. Epari, R. Steck, V. Glatt, I.C. Dickinson, P.F.M. Choong, M.A. Schuetz and D.W. Hutmacher, *Nature*, 2013, **3**, 216.
15. K.J. Burg, S. Porter and J.F. Kellam, *Biomaterials*, 2000, **21**, 23, 2347.
16. J.C. Middleton and A.J. Tipton, *Biomaterials*, 2000, **21**, 23, 2335.
17. B.N. Summers and S.M. Eisenstein, *Journal of Bone & Joint Surgery*, 1989, **71B**, 4, 677.
18. A. Baji, S.C. Wong, T.S. Srivatsan, G.O. Njus and G. Mathur, *Materials and Manufacturing Processes*, 2006, **21**, 20, 211.
19. K.A. Athanasiou, C.M. Agrawal, F.A. Barber and S.S. Burkhart, *Arthroscopy*, 1998, **14**, 7, 726.
20. J. Suganuma and H. Alexander, *Journal of Applied Biomaterials*, 1993, **4**, 1, 13.
21. R.R.M. Bos, F.R. Rozema, G. Boering, A.J. Nijenhuis, A.J. Pennings and H.W.B. Jansen, *British Journal of Oral and Maxillofacial Surgery*, 1989, **27**, 6, 467.
22. F. Namvar, M. Jawaid, P. Tahir, R. Mohamad, S. Azizi, A. Khodavandi, H.S. Rahman and M.D. Nayeri, *Bioresources*, 2014, **9**, 3, 19.
23. J.M. Anderson and M.S. Shive, *Advanced Drug Delivery Reviews*, 2012, **64**, SUPPL., 72.
24. C.E. Holy, J.A. Fialkov, J.E. Davies and M.S. Shoichet, *Journal of Biomedical Materials Research: Part A*, 2003, **65**, 4, 447.
25. J.G. Lyons in *Development of Novel Monolithic Matrices for Drug Delivery using Conventional and Non-conventional Polymer Processing Technologies*, Athlone Institute of Technology, Athlone, Ireland, 2007 [Doctoral Thesis].
26. P. Gunatillake, R. Mayadunne and R. Adhikari, *Biotechnology Annual Review*, 2006, **12**, 6, 301.
27. N. Angelova and D. Hunkeler, *Trends in Biotechnology*, 1999, **17**, 10, 409.
28. C-P. Jiang, J-R. Huang and M-F. Hsieh, *Rapid Prototyping Journal*, 2011, **17**, 4, 288.
29. F. Chen, T. Mao, K. Tao, S. Chen, G. Ding and X. Gu, *British Journal of Oral and Maxillofacial Surgery*, 2003, **41**, 4, 240.
30. K.J.L. Burg and D.E. Orr in *An Overview of Bioresorbable Materials*, Woodhead Publishing Ltd, Cambridge, UK, 2008.
31. R.E. Cameron and A. Kamvari-Moghaddam in *Synthetic Bioresorbable Polymers*. Woodhead Publishing Ltd, Cambridge, UK, 2012.
32. M. Generali, P.E. Dijkman and S.P. Hoerstrup, *European Medical Journal*, 2014, **July**, 91.
33. G. Narayanan, V.N. Vernekar, E.L. Kuyinu and C.T. Laurencin, *Advanced Drug Delivery Reviews*, 2016, **107**, 247.
34. G.O. Hofmann, *Clinical Materials*, 1992, **10**, 1–2, 75.
35. W.R. Gombotz and D.K. Pettit, *Bioconjugate Chemistry*, 1995, **6**, 4, 332.
36. S. Kehoe, X.F. Zhang and D. Boyd, *Injury*, 2012, **43**, 5, 553.
37. S. Guns, V. Mathot, J.A. Martens and G. Van Den Mooter, *European Journal of Pharmaceutics and Biopharmaceutics*, 2012, **81**, 3, 674.
38. J.G. Lyons, D.M. Devine, J.E. Kennedy, L.M. Geever, P. O'Sullivan and C.L. Higginbotham, *European Journal of Pharmaceutics and Biopharmaceutics*, 2006, **64**, 1, 75.

39. J.G. Lyons, H. Holehonnur, D.M. Devine, J.E. Kennedy, L.M. Geever, P. Blackie and C.L. Higginbotham, *Materials Chemistry and Physics*, 2007, **103**, 2–3, 419.
40. J.G. Lyons, M. Hallinan, J.E. Kennedy, D.M. Devine, L.M. Geever, P. Blackie and C.L. Higginbotham, *International Journal of Pharmaceutics*, 2007, **329**, 1–2, 62.
41. K. Kim, M. Yu, X. Zong, J. Chiu, D. Fang, Y.S. Seo, B.S. Hsiao, B. Chu and M. Hadjiargyrou, *Biomaterials*, 2003, **24**, 27, 4977.
42. A. Oryan, S. Alidadi, A. Moshiri and N. Maffulli, *Journal of Orthopaedic Surgery and Research*, 2014, **9**, 1, 18.
43. A.R. Boccaccini and V. Maquet, *Composites Science and Technology*, 2003, **63**, 16, 2417.
44. R. Scaffaro, F. Lopresti, L. Botta, S. Rigogliuso and G. Ghersi, *Interface Tissue Engineering*, 2016, **54**, 8.
45. S. Farah, D.G. Anderson and R. Langer, *Advanced Drug Delivery Reviews*, 2016, **107**, 367.
46. K. Fukushima, C. Abbate, D. Tabuani, M. Gennari and G. Camino, *Polymer Degradation and Stability*, 2009, **94**, 10, 1646.
47. Y. Chen, L.M. Geever, J.A. Killion, J.G. Lyons, C.L. Higginbotham and D.M. Devine, *Polymer Composites*, 2015, DOI:10.1002/pc.23794.
48. M. Labet and W. Thielemans, *Chemical Society Reviews*, 2009, **38**, 12, 3484.
49. B.D. Ulery, L.S. Nair and C.T. Laurencin, *Journal of Polymer Science, Part B: Polymer Physics*, 2011, **49**, 12, 832.
50. K. Van De Velde and P. Kiekens, *Polymer Testing*, 2002, **21**, 4, 433.
51. P. Gunatillake, R. Mayadunne and R. Adhikari, *Biotechnology Annual Review*, 2006, **12**, 301.
52. P.D. Darney, S.E. Monroe, C.M. Klaisle and A. Alvarado, *American Journal of Obstetrics and Gynecology*, 1989, **160**, 5 Pt 2, 1292.
53. M.A. Woodruff and D.W. Hutmacher, *Progress in Polymer Science (Oxford)*, 2010, **35**, 10, 1217.
54. Y. Tokiwa and T. Suzuki, *Nature*, 1977, **270**, 3, 76.
55. M. Forouharshad, L. Gardella, D. Furfaro, M. Galimberti and O. Monticelli, *European Polymer Journal*, 2015, **70**, 28.
56. J. Rodenas-Rochina, A. Vidaurre, I. Castilla Cortázar and M. Lebourg, *Polymer Degradation and Stability*, 2015, **119**, 121.
57. Z. Ning, N. Jiang and Z. Gan, *Polymer Degradation and Stability*, 2014, **107**, 120.
58. I. Navarro-Baena, V. Sessini, F. Dominici, L. Torre, J.M. Kenny and L. Peponi, *Polymer Degradation and Stability*, 2016, **132**, 97.
59. V.C.C. Trajano, K.J.R. Costa, C.R.M. Lanza, R.D. Sinisterra and M.E. Cortés, *Materials Science and Engineering: C*, 2016, **64**, 370.
60. N. Silveira, M.M. Longuinho, S.G. Leitão, R.S.F. Silva, M.C. Lourenço, P.E.A. Silva, M. doC.F.R. Pinto, L.G. Abraçado and P.V. Finotelli, *Materials Science and Engineering: C*, 2016, **58**, 458.
61. M.C.P. Brugmans, S.H.M. Söntjens, M.A.J. Cox, A. Nandakumar, A.W. Bosman, T. Mes, H.M. Janssen, C.V.C. Bouten, F.P.T. Baaijens and A. Driessen-Mol, *Acta Biomaterialia*, 2015, **27**, 21.
62. A. Göpferich, *Biomaterials*, 1996, **17**, 2, 103.
63. J. Bergsma, W.C. de Bruijn, F.R. Rozema, R.R.M. Bos and G. Boering, *Biomaterials*, 1995, **16**, 1, 25.
64. T-W. Chung, M-C. Yang, C-C. Tseng, S-H. Sheu, S-S. Wang, Y-Y. Huang and S-D. Chen, *Biomaterials*, 2011, **32**, 3, 734.
65. F. Rezgui, M. Swistek, J.M. Hiver, C. G'Sell and T. Sadoun, *Polymer*, 2005, **46**, 18, 7370.
66. R. Zheng, H. Duan, J. Xue, Y. Liu, B. Feng, S. Zhao, Y. Zhu, Y. Liu, A. He, W. Zhang, W. Liu, Y. Cao and G. Zhou, *Biomaterials*, 2014, **35**, 1, 152.
67. G.V. Salmoria, R.A. Paggi, F. Castro, C.R.M. Roesler, D. Moterle and L.A. Kanis, *Procedia CIRP*, 2016, **49**, 193.

68. R. Sartoneva, S. Haimi, S. Miettinen, B. Mannerström, A-M. Haaparanta, G.K. Sándor, M. Kellomäki, R. Suuronen and T. Lahdes-Vasama, *Journal of the Royal Society Interface*, 2011, **8**, 58, 671.
69. J. Wang, A. Goyanes, S. Gaisford and A.W. Basit, *International Journal of Pharmaceutics*, 2016, **503**, 1–2, 207.
70. A.S. Rodrigues, *Tissue Engineering*, 2011, **18**, 18, 225.
71. P.A. Gunatillake, R. Adhikari and N. Gadegaard, *European Cells and Materials*, 2003, **5**, 1.
72. P.B. Maurus and C.C. Kaeding, *Operative Techniques in Sports Medicine*, 2004, **12**, 3, 158.
73. V. Singh and M. Tiwari, *International Journal of Polymer Science*, 2010, **2010**, 1.
74. I. Vroman and L. Tighzert, *Materials*, 2009, **2**, 2, 307.
75. P. Gentile, V. Chiono, I. Carmagnola and P.V. Hatton, *International Journal of Molecular Sciences*, 2014, **15**, 3, 3640.
76. D.K. Gilding and A.M. Reed, *Polymer*, 1979, **20**, 12, 1459.
77. C.C. Chu, *Journal of Applied Polymer Science*, 1981, **26**, 5, 1727.
78. T.M. Reed, *The Journal of Foot and Ankle Surgery*, 1998, **38**, 1, 14.
79. W. Amass, A. Amass and B. Tighe, *Polymer International*, 1998, **47**, 2, 89.
80. R. Kronenthal, *Polymers in Medicine and Surgery*, 1975, **1**, 119.
81. D.F. Williams and E. Mort, *Journal of Bioengineering*, 1977, **1**, 3, 231.
82. R. Chandra and R. Rustgi, *Progress in Polymer Science*, 1998, **23**, 97, 1273.
83. H. Cao and N. Kuboyama, *Bone*, 2010, **46**, 2, 386.
84. R-J. Chung, K-L. Ou, W-K. Tseng and H-L. Liu, *Surface and Coatings Technology*, 2016, **303**, 283.
85. L. Cui, Y. Wu, L. Cen, H. Zhou, S. Yin, G. Liu, W. Liu and Y. Cao, *Biomaterials*, 2009, **30**, 14, 2683.
86. N. Mahmoudifar and P.M. Doran, *Biomaterials*, 2010, **31**, 14, 3858.
87. H.J.Z.R. Costa, R. Ferreira Bento, R. Salomone, D. Azzi-Nogueira, D.B. Zanatta, M. Paulino Costa, C.F. da Silva, B.E. Strauss and L.A. Haddad, *Brain Research*, 2013, **1510**, 10.
88. K. Chwalek, Y. Dening, C. Hinüber, H. Brünig, M. Nitschke and C. Werner, *Materials Science and Engineering: C*, 2016, **61**, 466.
89. C. Stoll, T. John, C. Conrad, A. Lohan, S. Hondke, W. Ertel, C. Kaps, M. Endres, M. Sittinger, J. Ringe and G. Schulze-Tanzil, *Biomaterials*, 2011, **32**, 21, 4806.
90. M.F. Pietschmann, B. Frankewycz, P. Schmitz, D. Docheva, B. Sievers, V. Jansson, M. Schieker and P.E. Muller, *Journal of Materials Science: Materials in Medicine*, 2013, **24**, 1, 211.
91. J. Cornah and J. Wallace, *British Journal of Oral and Maxillofacial Surgery*, 1988, **26**, 250.
92. H.R. Kricheldop and D. Damrau, *Macromolecules*, 1998, **199**, 1089.
93. H.S. Schultz, inventor; Gen Aniline & Film Corp, assignee; US3063967, 1962.
94. A. Kaplan, G. Ruff, J. Leung and M. Megaro, inventors; Ethicon, Inc., assignee; US8764796B2, 2014.
95. J-M. Raquez, P. Degée, R. Narayan and P. Dubois, *Macromolecular Rapid Communications*, 2000, **21**, 15, 1063.
96. A.K. Sugih, F. Picchioni and H.J. Heeres, *European Polymer Journal*, 2009, **45**, 1, 155.
97. G.Q. Chen and Q. Wu, *Biomaterials*, 2005, **26**, 33, 6565.

Yuanyuan Chen, Marcelo Jorge Cavalcanti De Sá, Maurice Dalton
and Declan M. Devine

2 Biodegradable medical implants

2.1 Introduction

Each year, the quality of life of millions of patients is improved through surgical
procedures that involve implanted medical devices. The US Food and Drug Administration (FDA) defines medical implants as 'devices or tissues that are placed inside
or on the surface of the body, replacing missing body parts, delivering medications,
monitoring body function and providing support to organs and tissues'. General
medical implants include orthopaedics, pacemakers, cardiovascular stents, defibril-
lators, neural prosthetics and/or drug-delivery systems [1].

Conventional materials such as metals, polymers and ceramics have found applications in various medical implants. Metals are widely used in load-bearing implants, ranging from plates and screws for bone fracture fixation to joint prostheses
for hips, knees, shoulders, and ankles. The most commonly employed metals are
grade 316L stainless steels, cobalt-chromium alloys, titanium alloys and magnesium alloys [2, 3]. Polymers have been used in facial prostheses, kidney and liver
parts, heart components, dentures, as well as hip and knee joints. For instance,
ultrahigh-molecular weight polyethylene has been employed for load-carrying devices [4], polymethyl methacrylate as bone cement [3], and polyethylene terephthalate (PET) for vascular grafts and heart valves [5]. Ceramics have been used to
replace or fix hard connective tissue, such as bone and teeth, due to their high
strength, toughness and surface finish [6].

Biodegradable materials have been utilised in medical applications and have
attracted a lot of attention. The development of fully biodegradable medical implants started with biodegradable sutures first approved in 1960s [6]. Biodegradable
implants provide a temporary support and degrade after service. The degradable
nature of biodegradable medical implants allows for excretion of the materials from
the body, enabling the injured site to restore its function over time after having
benefited from the implants [6]. There are four types of degradation mechanisms
responsible for implants degradation: hydrolysis, oxidation, enzymatic and physical degradation [7]. Polyhydroxy acids are the main materials for biodegradable
devices due to their good biocompatibility. These materials include polylactic acid
(PLA), polyglycolic acid (PGA), and copolymers based on lactide, glycolide, trimethylene carbonate (TMC) and polycaprolactone (PCL) [6].

Biodegradable implants offer many advantages over traditional permanent
implants, especially metallic implants, which are related to stress shielding, corro-
sion, ion leaching, inflammation, and implant removal. There has been a shift in
the use of the permanent metallic implants for temporary therapeutic applications

https://doi.org/10.1515/9783110640571-002

to biodegradable implants. This chapter mainly discusses the advance of biodegradable implants used in orthopaedic, cardiovascular and wound-closure systems.

2.2 Biodegradable sutures

Sutures have been widely used in wound closure for centuries. Collagen, cotton, silk, linen, polypropylene (PP), PET, polybutylene terephthalate (PBT), and polyamide have been used as suture materials. However, since the successful introduction of two synthetic absorbable sutures, Dexon® and Vicryl®, in the early 1970s, a new chapter of suture materials has been opened [8]. Biodegradable sutures can break down in tissue after the wound is closed firmly in a given period of time, and are used extensively for healing internal wounds to avoid secondary surgery to remove sutures [9]. Originally, biodegradable sutures were made from sheep intestines and called 'catgut'. Biodegradable sutures have numerous advantages over catgut sutures, including ease of handling, low cost, and low tissue reaction. Synthetic biodegradable polymers such as poly(lactic-*co*-glycolic acid) (PLGA), PGA, PLA, polyglyconate, and PLA-*co*-ε-caprolactone (ε-CL) are currently the main materials for biodegradable sutures [10]. Common commercial biodegradable sutures and their polymeric materials are listed in Table 2.1.

Table 2.1: Common commercial biodegradable sutures.

Suture name	Material	Manufacturer
Dexon®	PGA	Davis & Geck Corporation
Vicryl®	PLGA	Ethicon
Maxon®	Polyglyconate	Davis & Geck Corporation
Monocryl®	PGA-*co*-ε-caprolactone	Ethicon
Biosyn®	Polydioxinone-*co*-trimethylene carbonate-*co*-glycolide	Formerly US Surgical
Bio-Anchor	PLLA	Mitek
Vicryl Rapide® (Polyglactin 910)	Glycolide/PLLA	Ethicon
Bio-PushLock	PLLA	Arthrex
Polysorb	PLLA/PGA	Covidien-Medtronic
Vicryl Plus®	Glycolide/PLLA with Triclosan	Ethicon

PLLA: Poly(L-lactic acid)
Reproduced with permission from Y. Chen, L.M. Geever, J.A. Killion, J.G. Lyons, C.L. Higginbotham and D.M. Devine, *Polymer-Plastics Technology and Engineering*, 2016, **55**, 10. ©2016, Taylor & Francis [11]

Biodegradable sutures are defined by the loss of their strength within 60 days after placement [12]. In most cases, 3 weeks is enough for a wound to close, and biodegradable sutures can keep tissue from separating for this period and degrade afterwards. Vicryl® is made of PLGA and retains 65% of its tensile strength at 2 weeks and 40% at 3 weeks. Complete absorption occurs between 60 days and 90 days by hydrolysis [13]. Dexon® is made of PGA and retains 89% of its tensile strength at 7 days, 63% at 14 days and 17% at 21 days [14]. Compared with Dexon®, Vicryl® shows a slower loss of function and higher knot-breaking strength, but less irreversible elongation [15].

If a wound requires <3 weeks to heal, fast-degrading sutures can be used. Vicryl Rapide® is made from the same material as Vicryl® but is partially hydrolysed in buffer solution and sterilised with gamma radiation. Hence, the degradation process is rapid and 50% of tensile strength is retained at 5 days, and by the 2 weeks the tensile strength is 0%. Monocryl® is made of PGA-*co*-ε-CL and retains 25–40% of its tensile strength at 2 weeks and 0% at 21 days [13]. Compared with Vicryl Rapide®, Monocryl® can result in significantly smaller and less reactive scars, and a lower tendency to form hypertrophic scars [16].

In some cases, a much longer healing period than 3 weeks is required. In these instances, polydioxanone (PDO) retains 74% of its tensile strength after 2 weeks, 50% after 4 week and 25% after 6 weeks. PDO is stiff and difficult to handle, and is a low-reactivity suture that maintains its integrity in the presence of bacterial infection. Maxon™ is made of polyglyconate, retains its tensile strength up to 92 days, and absorption is complete in 6–7 months [13]. PLA sutures have more prolonged retention of tensile strength than that of Maxon™ and they have been suggested as alternatives in repair of the Achilles tendon [17]. Self-reinforced PLA sutures exhibit the most prolonged retention of tensile strength and can be applied to the closure of wounds that need prolonged support, such as bone repair [18]. Recently, a drawn PLLA suture with a helical structure was investigated in rat patellar ligaments and reported to generate a piezoelectric charge under cyclically applied tensile stress. Significantly higher ossification was observed around the implanted helical PLLA suture in the rat knee joints compared with untwisted PLLA sutures, which suggested that helical PLLA fibres may be useful for surgical sutures or artificial ligaments connecting to the bone [19].

Antibacterial sutures are a recent development for wound closure. Surgical-site infections are challenging complications, and antimicrobial-coated sutures can defend against various bacterial pathogens effectively [20]. Vircryl Plus® is made of 90% glycolide and 10% L-lactide, and produced by Ethicon. This suture contains Irgacare® MP (troclosan), a broad-spectrum antibacterial agent. The first instances of triclosan resistance were reported recently. Hence, an alternative chlorhexidine was coated on sutures by Obermeier and co-workers and demonstrated high antimicrobial efficacy against *Staphylococcus aureus in vitro* [20].

Use of fibre mats to cover wounds is another recent development. It is a very promising strategy if the wound shape is complex. Biodegradable polymeric fibres can be fabricated into mats of various shapes and used for medical treatment. Behrens and co-workers reported the use of solution blow spinning to generate PLGA nanofibres on the wound surface *in situ* utilising a commercial airbrush and compressed carbon dioxide (CO_2) [21]. The PLGA fibres airbrushed onto the incision of the wound can take the shape of the area and promote healing [22]. Zhang and co-workers developed multilayered PLA electrospun nanofibres loaded with cisplatin. They covered the surgical site with this PLA mat following resection of subcutaneous liver cancer in mice, and reported less tumour recurrence, prolonged survival time of mice, and less systemic toxicity compared with other treatment groups [23]. Similarly, Zhang and co-workers implanted PLA nanofiborous mats loaded with 5-fluorouracil and oxaliplatin into mice with colorectal cancer, and found a suppressed tumour growth rate and prolonged survival time of mice in comparison with a drug-free control group [24].

Biodegradable sutures have replaced traditional non-biodegradable sutures in most cases. Biodegradable sutures can keep the wound tissue closed for a short period and break down in tissue afterwards, limiting the risk of surgical-site infection and avoiding secondary surgery to remove the suture. Non-biodegradable sutures are still used in cases that require a suture to stay for a long time (e.g., heart and blood vessels with rhythmic movements) and organs (e.g., bladder) containing fluids that speed-up the degradation process of biodegradable sutures. However, with recent developments, such as reinforced PLA sutures used in bone repair, the use of biodegradable sutures will continue to widen.

2.3 Bone-fixation devices

Bone fracture results in a completely detached or partially attached bone fragments which, without fixation results in imperfect healing by cartilaginous tissue [25]. Therefore, it is common to use fixation devices such as plates, screws, pins, nails and rods to hold the bone fragments together to allow regrowth and healing. Such devices are traditionally constructed of metals, which have been reported to cause medical complications [26].

Stress shielding is one of the major complications related to metal fixation devices [27]. It occurs if two or more components with different moduli form a single mechanical system. The component with the higher modulus bears most of the load and protects the other component from stress. Such is the case with rigid metallic fixation devices, which at first may favour primary bone healing with sufficient mechanical support. However, at the later stages of healing, stress shielding occurs and the bone recovers with insufficient strength, which can lead to osteoporosis [28]. Thus, the bone may have a tendency to fracture again if the metal plates are removed [29]. In addition, metallic bone-fixation devices have been reported to interfere with skeletal growth, particularly for paediatric patients and magnetic resonance imaging (MRI) investigations [30–36]. Moreover, leaching of metal ions to adjacent tissue is problematic. For instance, titanium as a common metal used for bone fixation but has been found in the lymph nodes, liver, spleen, bone marrow, and brain [35, 36]. Finally, implant removal for functional improvement and pain relief, which accounts for 15% of all surgical procedures in the orthopaedic and trauma units in Scandinavia, is expensive and challenging [37].

Fully biodegradable bone-fixation devices appear to be great alternatives for patients. The major benefits of fully biodegradable bone-fixation devices include no requirement for implant removal, minimal risk of implant-related complications, and early functional rehabilitation [38]. Fully biodegradable bone-fixation devices are often made of biodegradable polymers such as PLA, PGA, PLGA, PDO, polysulfate and polycarbonate (PC) [39]. PLA is a relatively strong biodegradable polymer and has been used intensively in medical applications. With L- and D-enantiomeric forms of lactic acid, PLA appears as PLLA with only the L-isomer, poly(D-lactic acid) (PDLA) with only the D-isomer, poly(D,L-lactic acid) (PDLLA) with half L-isomer and half D-isomer, and PLA with a majority of the L-isomer and a small amount of the D-isomer [40]. Commonly used biodegradable bone-fixation devices and their materials are listed in Table 2.2.

Table 2.2: Commonly used biodegradable bone fixation devices.

Devices	Material	Manufacturer
SmartNail	PLLA	Conmed Linvatec
LactoNail	PGA/PLLA	Arthrotek Biomet
SmartScrew	PLLA	Conmed Linvatec
ReUnite Screw	PGA/PLLA	Arthrotek Biomet
BioComposite	Biphasic calcium/PDLLA	Arthrex
Hexalon™	L–lactide/D-lactide/TMC	Inion Limited
BioCore	PLLA/PGA	Biomet
The wedge	Self-reinforced PDLLA	ConMed Linvatec

Reproduced with permission from Y. Chen, L.M. Geever, J.A. Killion, J.G. Lyons, C.L. Higginbotham and D.M. Devine, *Polymer-Plastics Technology and Engineering*, 2016, **55**, 10. ©2016, Taylor & Francis [11]

Biodegradable bone-fixation devices have an advantage over traditional rigid metallic devices because implant removal is not required, which is especially beneficial in spine repair. Loosening of permanent implants is a serious complication requiring subsequent surgical removal of these implants [41, 42]. Removal of permanent implants in the spine due to instability or dislocation may leave a deficit leading to instability of the spine [43, 44]. Therefore, biodegradable fixation devices, which provide mechanical support and degrade with the pace of the healing process, offer an ideal option for spine fixation. Spinal biodegradable implants include a myriad of posterior lumbar interbody fusion devices, anterior spinal plates as well as screws and meshes [45].

Furthermore, biodegradable bone-fixation devices have been found to be superior to traditional metallic devices in terms of biocompatibility and functional recovery. Research has been conducted to compare tissue reactions to titanium and biodegradable screws upon postoperative irradiation in rats. It was found that the intensity of inflammation around the titanium screws was more evident than that around the biodegradable screws, which indicated that biodegradable screws could be safer reconstructive devices than titanium screws in patients undergoing radiotherapy [46]. Several studies have shown the rate of infection and plate removal of biodegradable plates to be significantly lower than that of titanium plates [47, 48]. A follow-up study of 10 patients with maxillofacial fractures who had undergone implantation with biodegradable Inion CPS® reported favourable healing and no local tissue immune reaction [49]. In addition, an Inion CPS® biodegradable fixation system was evaluated in 8 patients with fractures of the mandibular body, and sufficient mechanical strength and stable fixation were observed [50]. Inion CPS® was reported to be a reliable fixation device which elicited only minor infections in 19 patients with a unilateral fracture of the mandibular angle [51]. Biodegradable fixation devices have adequate shear resistance, but less load resistance and stiffness compared with titanium, which helps to avoid stress shielding and promotes healing [52]. The use of biodegradable fixation devices is a promising technology and good healing can be achieved in patients of all ages, as well as mandible fractures in early childhood [25, 49].

The average Young's modulus of trabecular bone and cortical bone has been measured to be 10.4 and 18.6 GPa, respectively [53], and the tensile strength of bones has been reported to be 50–150 MPa [54, 55]. Biodegradable polymers do not have comparable mechanical strength, and are not suitable for many orthopaedic applications [56]. For instance, the tensile yield strength of PGA and PCL has been reported to be 57 and 19–21 MPa, respectively [57]. PLA is one of the strongest biodegradable polymers with tensile modulus of 3–4 GPa, tensile strength of 50–70 MPa and elongation at break of 2–10% depending on molecular weight (MW), crystallinity and processing methods [56, 58]. Therefore, biodegradable polymers reinforced with various fillers have been studied intensively. Hydroxyapatite (HA) nanorod-reinforced PLLA composites were found to have a Young's modulus of 1.3 GPa and a compressive strength of 110.3 MPa [59]. OSTEOTRANS-MX® is made by Takiron from unsintered HA–PLLA composites and has clinically proved to be beneficial in treatment of maxillofacial fractures (Figure 2.1). Basalt fibre-reinforced PLA composites were developed and had a Young's modulus of 8.31 GPa and a tensile strength of 123.5 MPa with 40 wt% loading of basalt fibre [60]. Self-reinforced PLA was found to have a Young's modulus of 3.3 GPa and a tensile strength of 48

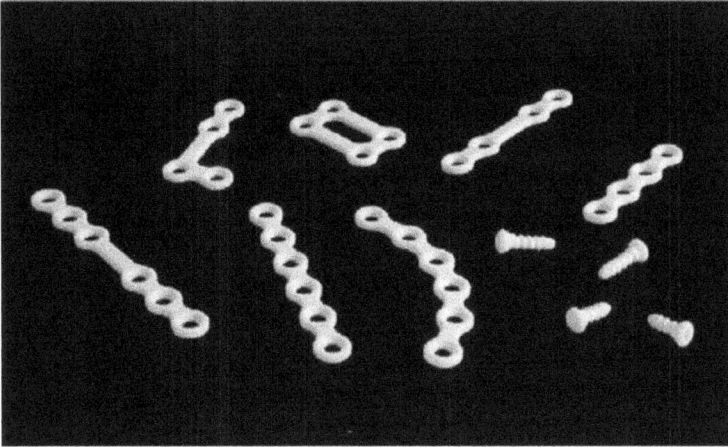

Figure 2.1: OSTEOTRANS-MX® composed of a forged unsintered HA–PLA composite. Reproduced with permission from S. Sukegawa, T. Kanno, N. Katase, A. Shibata, Y. Takahashi and Y. Furuki, *The Journal of Craniofacial Surgery*, 2016, **27**, 1391. ©2016, Mutaz B. Habal, MD [62].

MPa [27]. Phosphate glass fibre-reinforced PLA was reported to have a flexural strength of 110–180 MPa and a modulus of 11–18 GPa, which might be beneficial for high loading applications [61]. Polyvinyl alcohol (PVA)–HA composite reinforced with catgut fibres was found to have adequate mechanical strength for fixation of the facial skeleton with a tensile strength of 39.88 ± 1.33 MPa after a degradation period of 60 days when compared with the mechanical strength of the masseter muscle in the molar region (14.28 MPa) [55]. Reinforcing fillers can improve the mechanical properties of biodegradable polymers. The adhesion between the polymer matrix and fillers determines the final mechanical properties of the composites. A desirable interfacial affinity between the polymer matrix and fillers can optimise enhancement of the mechanical properties of composites [62].

The ideal biodegradable bone-fixation devices not only provide physical support, but also degrade at the pace of the healing process. Degradation of bone plates and screws made from PDLLA have a long degradation period (≈2 years), and devices made from self-reinforced PLLA take up to 5–6 years to be resorbed completely, whereas those made from PGA can be resorbed in 3 months and are suitable for short-term bone fixation [57]. The degradation of these implants has been postulated to be the cause of marked foreign-body reactions, synovitis, and even activation of the complement cascade [63]. The foreign-body reactions of PLA-based implants have been reported. PDLLA plates with marked local foreign body reactions have been documented as early as 3 months after implantation. However, PLA plates, screws and rods have been reported to cause significantly less local soft-tissue reactions compared with PGA devices [64–66]. The rapid degradation of PGA is the cause for the soft-tissue reactions [63]. By changing the copolymer ratios of PGA and PLA, the degradation rate can be optimised [63]. Foreign-body reactions have not been reported with PLGA plates [67].

In addition, biodegradable polymeric composites have been reported to promote bone healing, and implants made from these materials are very promising. PDLLA-*co*-TMC and nano-HA composites have been found to enhance alkaline phosphatase (ALP) secretion as well as mineral deposition during bone formation. Also, the feasibility of using them for bone healing after drilling of screw holes is currently under investigation [68]. Zhang and co-workers developed a PLA/ demineralised bone matrix (DBM) composite and found that PLA combined with DBM promoted better bone healing than PLA alone in the radial bones of rabbits [69].

The benefits of biodegradable fixation devices outweigh those of traditional bone-fixation devices because:

1. Biodegradable plates, rods and screws do not interfere with imaging and allow a clear view of fracture sites on radiographs [48, 70].
2. Biodegradable plates are easier to handle than metallic plates, and bending instruments such as pliers are not needed to place biodegradable plates against bone surfaces [47].
3. No need for a procedure to remove them, which reduces costs [70].
4. Biodegradable bone-fixation devices promote better bone healing than traditional plates and screws [46, 48–51].

However, the problems related to biodegradable plates and screws have also been reported:

1. Biodegradable screws require a drill hole that must be tapped, and this increases the operative time [71].
2. It is more difficult to place biodegradable screws than metallic screws [72]. Biodegradable screws require slower drilling rates into bone [71]. The high drilling speed might cause heat due to friction, which can cause difficulties when placing the screws [73].
3. Polymer debris causing inflammatory foreign-body reactions has been reported [74].

2.4 Fully biodegradable stents

Percutaneous coronary intervention (PCI) is the standard procedure for treating coronary heart disease (CHD). PCI started with balloon angioplasty, but it had a very high rate of restenosis, recoil and constrictive remodelling. Therefore, balloon angioplasty has been replaced by stenting [75]. Bare metallic stents (BMS) were revolutionary inventions but, due to the major problem of in-stent restenosis, drug-eluting stents (DES) have taken their place [76]. DES coated with antiproliferative drugs can effectively reduce the rate of in-stent restenosis, but late stent restenosis has being a recent cause for concern [77]. The lumen size after DES implantation remains fixed over the long term and is associated with clinical incidences, including death, myocardial infection MI, coronary artery bypass grafting CABG, target lesion revascularisation and so on [78, 79]. Neointimal tissue growth results in restenosis requiring an additional intervention in about 2–5% of DES cases [80]. Fully biodegradable coronary stents provide transient vessel support with drug-delivery capability and degrade after service, providing temporary support without the long-term limitation of metallic stents, and appear to be the ideal option to treat CHD [81].

Efforts to create biodegradable stents started approximately 20 years ago. However, the common types of fully biodegradable materials, including PLGA, PCL, poly (hydroxybutyrate-*co*-hydroxyvalerate) copolymer, polyorthoester, polyethylene oxide (PEO)/PBT and low-MW PLLA (80 kD), resulted in intense inflammation leading to neointimal hyperplasia and/or thrombus formation. This technology failed to develop due to inflammation, restenosis and growing interest in DES [82, 83]. The capability of producing high-MW PLA and biodegradable metals such as magnesium brought fully biodegradable stents back into the research field. Currently, 28 companies are developing fully biodegradable stents. The most commonly used biodegradable material is PLLA (24 products) due to its relatively high strength, excellent biodegradability and good biocompatibility [84]. It has a long history in the stents industry as a polymer coating and drug carrier for DES, which helps to streamline regulatory approval [85]. Other materials being used are tyrosine PC, salicylic acid polymer, magnesium and iron [81]. Several fully biodegradable polymeric stents that are currently being manufactured are presented in Table 2.3 and Figure 2.2.

Table 2.3: Fully biodegradable polymeric coronary stents [90–93].

Stent	Manufacturer	Material
Absorb™ BVS	Abbott	PLLA
Igaki-Tamai	Kyoto Medical	PLLA
DESolve Nx	Elixir®	PLLA
ReZolve	Reva	Tyrosine PC
MeRes100	Meril Life Sciences	PLLA
Ideal	Bioabsorbable Therapeutics Incorporated	Polyanhydride ester salicylic acid

The Igaki–Tamai stent was the pioneering fully biodegradable stent made from PLLA and it is currently used in peripheral arteries. Data from 50 patients over 10 years showed that some acute recoil of the stent occurred. However, this stabilised quickly and the performance at 6-month follow-up was satisfactory with initial

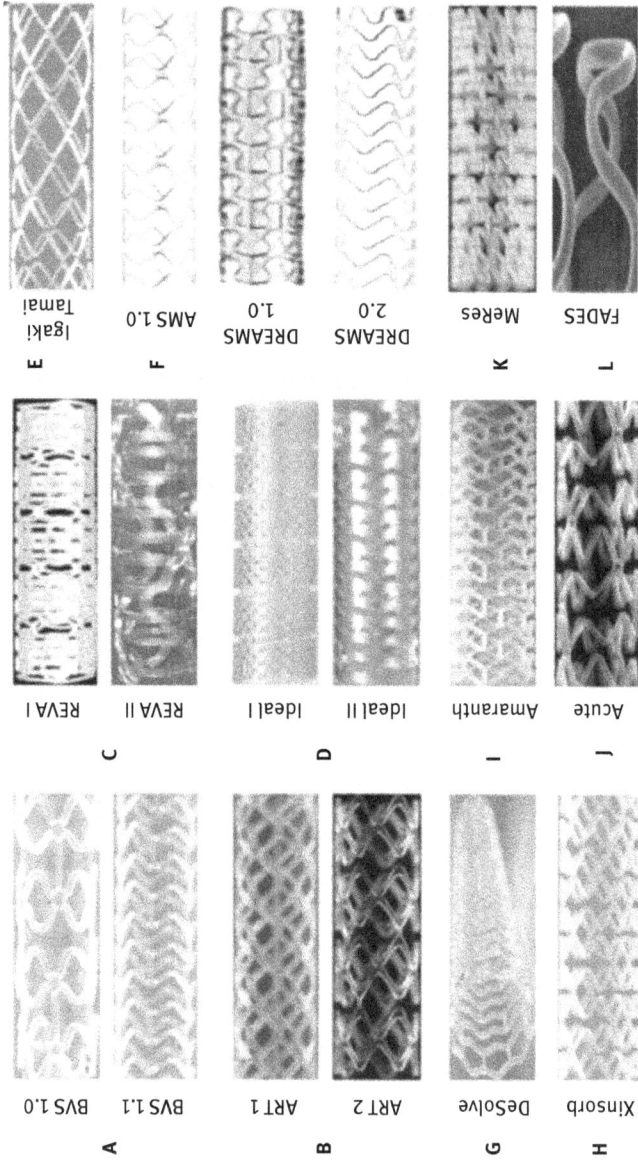

Figure 2.2: Currently available fully biodegradable stents. **A)** Abbott Vascular 1.0, 1.1; **B)** ART 1, 2; **C)** REVA Medical I, II; **D)** Ideal scaffold I, II; **E)** Igaki–Tamai scaffold; **F)** magnesium absorbable stents; **G)** DeSolve™; **H)** Xinsorb; **I)** Amaranth; **J)** Acute; **K)** MeRes; and **(L)** FADES. Reproduced with permission from Y. Zhang, C.V Bourantas, V. Farooq, T. Muramatsu, R. Diletti, Y. Onuma, H.M. Garcia-Garcia and P.W. Serruys, *Medical Devices: Evidence and Research*, 2013, **6**, 37. ©2013, Dove Press Ltd [90].

hyperplasia comparable with that of BMS. Furthermore, a 4-year follow-up study showed the stents degraded completely with no further development of hyperplasia [91]. The Igaki–Tmai stent has zigzag helical coils with a straight-bridge design and 24% stent-to-artery coverage ratio, and the strut thickness is 170 μm [92]. This stent is self-expandable, and requires heat at 70 °C to expand. The full deployment takes 20–30 min and due to the complicated delivery methods requiring heat for self-expansion, Igaki-Tmai is not currently used in coronary arteries [88].

Biodegradable vascular scaffold (BVS) everolimus-eluting PLLA coronary stents are manufactured by Abbott. They are balloon-expandable stents with zigzag hoops. The strut thickness is 158 μm, the crossing file is 1.4 mm, and stent-to artery coverage is 25% [92]. Zero stent thrombosis was documented from a clinical trial of 130 patients [87]. The FDA approved the Absorb™ BVS, making it the first commercially available biodegradable coronary stent [93]. However, very recent feedback of Absorb™ BVS use from cardiologists conducted by the Cardiovascular Research Foundation revealed it to be very promising but cost was one of the main issues. Absorb™ BVS costs approximately ten-times that of a standard metallic DES, and the potential advantages of Absorb™ BVS would not be expected until 2, 3, or even 5 years from now. Moreover, when dealing with Absorb™ BVS, use of online quantitative angiography, intravascular ultrasound, or optimal coherence tomography is strongly recommended to correctly measure and confirm vessel sizing, which adds to the procedure costs of Absorb™ BVS [93].

Another major issue is the increased risk of scaffold thrombosis due to a thicker strut of biodegradable polymeric stents compared with that of metallic stents. For instance, Absorb™ BVS has a strut thickness of 158 μm [92], whereas Synergy™ (which is made from a platinum chromium alloy) has a strut thickness of 74 μm. Biodegradable polymers are weak, so the stents made from biodegradable polymers have to be bulky to withstand the pressure from the vessel wall. The strut thickness directly links to stent performance and restenosis rate. Sorin Brener (New York Methodist Hospital, Brooklyn, NY, USA), called Absorb™ BVS 'clunky' and that it reminded him of the Cypher-type of first-generation DES. This safety concern was observed in the pivotal ABSORB III trial, which found the rate of definite/probable device thrombosis to be numerically higher with Absorb™ BVS, although the difference was not significant compared with the DES Xience® (1.5% *versus* 0.7%; p = 0.13). In addition, physicians have found that they have to dilate the vessel more when implanting biodegradable stents because they do not have radial strength whereas, with DES, even if the lesion is not fully open, good results are achievable [93]. Reducing the strut thickness of the stents made from biodegradable polymers may be achieved by reinforcing biodegradable polymers. Halloysite nanotube (HNT)-reinforced PLA composites were found to have a Young's modulus of 2 GPa and a tensile strength of 62.69 ± 4.25 MPa with 5 wt% loading of HNT. For further improvement in mechanical strength, surface treatment of HNT is under investigation [94].

Furthermore, the traditional method of laser machining, which has been employed for the manufacture of metallic stents, has also been used to fabricate biodegradable stents, such as Absorb™ BVS, but the problems of the heat-affected zone and solidification of molten material on cut edges have been reported [95]. Improvements in the manufacture of biodegradable polymeric stents have been investigated. Femtosecond lasers were used by Heublein and co-workers to fabricate biodegradable stents but high cost was an issue [96]. Excimer lasers were used by Barenghi and co-workers to fabricate biodegradable polypropylene fumarate (PPF) stents [97]. Stepak and co-workers utilised a CO_2 laser and claimed that elements cut with a CO_2 laser have better mechanical properties than those fabricated with excimer lasers [95]. Apart from laser machining, other fabrication methods include rapid prototyping [98], weft-knitting [99] and braiding technology [100].

Fully biodegradable polymeric stents are very promising, but they are early-stage technologies. Currently, only the Absorb™ and Elixir® scaffolds have a Conformité Européenne mark, and other scaffolds are undergoing clinical trials [81]. The advantages of fully biodegradable stents include:

1. No requirement for surgical removal due to the degradation of fully biodegradable stents.
2. Improved recovery of blood vessels compared with DES and BMS because fully biodegradable stents do not persist long-term and trigger late-stage thrombosis.
3. Reduction in the prevalence of bleeding complications because there is no requirement for long-term dual antiplatelet therapy.
4. Improvement in the treatment of complex multi-vessel disease that frequently results in the use of multiple long DES. Previously implanted biodegradable stents are not an issue if revascularisation is required in complex cases [101].
5. Polymeric stents allow the use of non-invasive imaging methods such as computed tomography and MRI for follow-up, whereas metallic stents can cause a 'blooming effect', making interpretation difficult [102].

Various challenges of fully biodegradable stents remain:

1. Fully biodegradable stents are very expensive compared with traditional metallic stents.
2. The clinical-trial experience of fully biodegradable stents is limited, and long-term follow-up is required.
3. Fully biodegradable polymeric stents have lower radial strength and rigidity compared with metallic stents, which contributes to a high rate of recoil [103].
4. The strut of fully biodegradable polymeric stents is much thicker than that of metallic stents, which is directly linked to stent performance and restenosis rate [104].
5. The polymers that have been used to develop biodegradable stents (e.g., PLLA, PLGA and PLLA-based polyester block copolymers) exhibit a brittle fracture mechanism at physiological conditions (37 °C), in which there is little or no

plastic deformation prior to failure. Thus, these polymeric struts can crack or fracture during crimping, delivery and deployment [103].

6. Some polymeric stents (e.g., BVS) have an approximate 2-year degradation period, which might be too long [105]. For the first 6 months, the stents are required to keep the vessel open while the vessel is healing and remodelling. After 6 months, the stents are required not to interfere with luminal enlargement of blood vessels, which often takes place between 6 months and 5 years after angioplasty [86].

2.5 Biodegradable anti-adhesive tissue barriers

Tissue adhesion following surgery is one of the most common challenges in clinical practice, particularly abdominal and bowel surgeries. Abdominal adhesions are not only the leading cause of small-bowel obstruction, but also cause at ≥20% of infertility cases and ≈40% of cases of chronic abdominal and pelvic pain. The adhesion formation typically occurs as a result of a fibrin clot, which transforms into scar tissue and connects different tissues that are normally separated in response to trauma, infection, dehydration, ischaemia, haemorrhage and foreign bodies [106, 107]. Surgical intervention is frequently required to eliminate the adhesions.

Multiple strategies to reduce adhesions include fibrinolytic agents, anti-inflammatory drugs and anti-adhesive barriers [108]. Recently, tissue adhesion barriers that cover tightly onto the damaged surface have been shown to physically isolate wounds, thus preventing the formation of tissue adhesions. Tissue adhesion barriers have become a hot topic in research and industrial fields. In the initial stages of developing tissue adhesion barriers, non-biodegradable polymers were used, such as polyethylene (PE) and expanded polytetrafluoroethylene. However, with the development of biodegradable barriers, traditional barriers, which require secondary surgical removal, have been gradually replaced. 'Ideal' biodegradable tissue anti-adhesive barriers should have good biocompatibility, biodegradability, immunological inertness, and do not interfere in wound healing [107].

Currently, biodegradable polymers such as PVA, PLA, PLA-polyethylene glycol (PEG), PLA-poly(oxyethylene-co-oxypropylene) (PLA-Pluronic F68), D,L-polylactide-ε-CL-poly(TMC) copolymer, hyaluronic acid, collagen, gelatin, silk, HA/carboxymethyl cellulose (CMC), and cellulose have been used extensively in anti-adhesive barriers [107]. PLA film has anti-adhesive properties due to low permeability against serum proteins as well as cells. PLLA nanosheets have been reported to not only seal a surgical incision without scarring, but also significantly reduce adhesion by inhibiting the permeation of oozing blood cells and infiltration of fibroblastic cells [109]. PLLA nanosheets have been found to reduce the tissue adhesion that accompanies liver injury [110].

Bilateral PLA/alginate membranes have been developed with a smooth PLA side keeping the affected tissues separated and a mucoadhesive side made of alginate to keep the barrier on the site of injury [111]. PLA nanosheets have been found to be better in reducing postoperative intestinal adhesion compared with Seprafilm®, which is made of HA and CMC, because PLA nanosheets did not affect bacterial propagation in the peritoneal cavity, whereas Seprafilm® showed bacterial propagation and led to increase mortality [112]. PLA pericardial membranes have been found to be promising for cardiac surgery [113]. Compared with PLA film, PVA-g-PLA film was found to have improved hydrophilicity and flexibility, and demonstrated good anti-tissue adhesion and suitable degradability [114]. Biodegradable PLGA electrospun membranes have been fabricated and found to be useful for anti-adhesion of the Achilles tendon [115].

Chitosan (CS) has been found to have anti-adhesive abilities when used as a coating on PP mesh [116]. CS–PLGA–PEO nanofibre mats were manufactured via electrospinning and showed good outcomes in craniofacial surgery [117]. Similar results have been reported with CS–PLGA nanofibrous sheets [118]. An injectable CS–dextran-based hydrogel has been developed as an adhesion-prevention postoperative aid, and claimed to have appropriate biocompatibility with a mild inflammatory response [119]. A PCL-gelatin hybrid membrane has been developed via electrospinning and revealed to possess good biocompatibility and degradation behaviour with anti-adhesion function in a rabbit model [120]. PEG–PCL nanofibrous membranes have been found to prevent cell penetration better than PCL membrane and Seprafilm® [121].

A novel hydrogel of PEG–PCL–PEG has been developed by Yang and co-workers. They reported that PEG–PCL–PEG hydrogel can adhere easily to damaged peritoneal surfaces and degrade within 14 days along with the healing of peritoneal wounds. This hydrogel has been found to prevent postoperative abdominal adhesions in mouse and rat models (Figure 2.3) [122].

There are various types of commercially available anti-adhesive barriers: Seprafilm® (HA/CMC), SurgiWrap® (PLA), Oxiplex® (PEG/ CMC), REPEL-CV® (PLA/PEG), Incert® (crosslinked HA), Interceed® (oxidised regenerated cellulose), Adept®, Intercoat®, SprayGel® and Guardix® [107, 108]. Most focus on the mechanical barrier by creating an inert barrier to cellular adhesion. However, some challenges have been reported which include: difficulty in handling sheet barriers in an operating room remains a problem; and the need for fixation using sutures which can lead to additional tissue adhesion [123]. Barriers made from crosslinked polymers have been reported to lose tissue adhesiveness because of a decrease in the number of specific functional groups interacting with tissues after the crosslinking procedure [107]. Despite of the limitations of biodegradable tissue barriers, an anti-adhesion approach utilising biodegradable polymers remains promising.

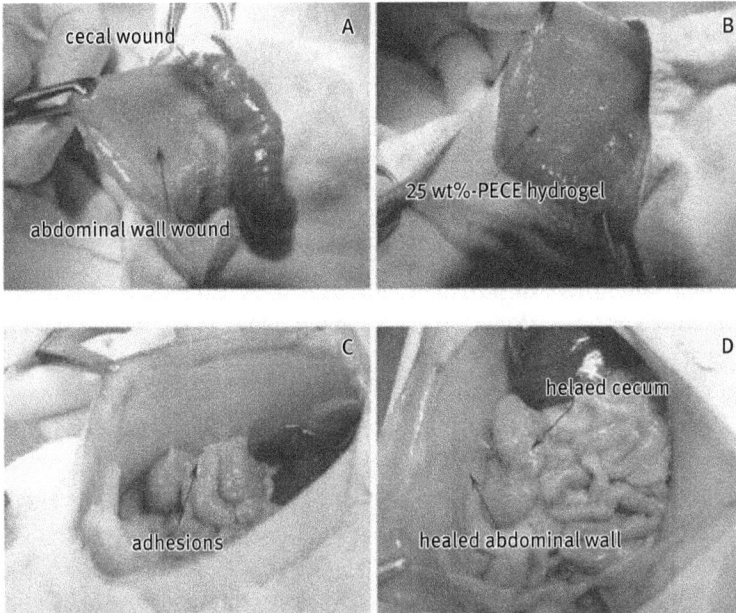

Figure 2.3: Prevention of abdominal adhesions in an abrasion model in rats. A) The establishment of a rat model of abdominal sidewall defect-cecum abrasion; B) poly(ε-caprolactone)– polyethylene glycol–poly(ε-caprolactone) (PECE) hydrogel applied on the injured abdominal wall and cecum; C) adhesion was observed in a saline-treated rat; and D) no apparent adhesion was observed in a PECE gel-treated rat, with a healed abdominal wall and cecum. Reproduced with permission from B. Yang, C. Gong, Z. Qian, X. Zhao, Z. Li, X. Qi, S. Zhou, Q. Zhong, F. Luo and Y. Wei, *BMC Biotechnology*, 2010, **10**, 65. ©2010, BioMed Central Ltd [122].

2.6 Tissue engineering

'Tissue engineering' was first defined at a National Science Foundation workshop in 1988 and defined further to the broad scientific community in 1993 by Langer and Vacanti [124]. Tissue engineering is a multidisciplinary field that aims to replace damaged tissues or organs. It does this by exploiting the natural physiological nature of an *in vivo* environment to temporarily replace damaged tissue. It utilises cells, growth factors, and implantable medical devices to aid patient recovery. In doing so, it focuses on the development, application and knowledge of chemistry, physics, engineering and life science [125]. This collaboration has led to immense research and rapid expansion in biomedical research due to its vast potential in the repair or replacement of tissue and organs [126]. Over recent years, materials science in conjunction with biotechnology has prioritised the development of artificial bone, organ implants, drug delivery and scaffolds for the engineering of bone, cartilage, vascular and nerve tissue [127]. The major aim of tissue engineering is to construct scaffolds that are capable of recreating an *in vivo* environment with the incorporation of biophysical, biomechanical and biochemical cues to aid in the cellular proliferation, differentiation and function [128]. For a scaffold to replace the mechanical function of living tissue, it must meet complex functional demands (pore shape, size and interconnectivity) to temporally generate an optimal environment for the generation process. For example, Ghorbani and co-workers constructed HA–gelatin scaffolds loaded with PLGA *via* freeze-casting for bone regeneration. Their study suggested that, by controlling the freeze rate, the porosity of the scaffold can be controlled. Nevertheless, lower freeze gradients compromised the mechanical features of the scaffold. They also explained that an increased concentration of HA can elicit beneficial mechanical properties to the scaffold [129].

Two methodologies are used in medical science for the regeneration of damage tissue or organ failure: transplantation and implantation. Transplants use donor tissue or organs (e.g., blood vessels or nerves) or cadavers (e.g., lyophilised and frozen bone) to aid the regeneration process. However, transplants have disadvantages such as ethical concerns and the requirement of immunosuppressive drugs to prevent rejection of the transplant. Implant devices are designed to act at the interface of recipient tissue. In addition, implant devices do not present as many issues in comparison with transplants due to the use of biomaterials [130]. Repair of bone and peripheral nerves *via* biomaterials are discussed in this section.

2.6.1 Bone tissue engineering

Traditionally, bone grafts such as autografts and allografts have been used to restore damaged bone, but both have limitations. Synthetic bone scaffolds used as bone graft substitutes have high mechanical strength and can enhance tissue growth [131]. Major advances in bone grafting have been achieved through the utilisation of growth factors, drugs and genes embedded within bone scaffolds. The ideal bone scaffolds should be three-dimensional (3D) and contain a highly porous network of interconnected pores to promote cell growth, adhesion, proliferation and differentiation by facilitating the flow and transport of nutrients and metabolic waste. In addition, they should have good biocompatibility and biodegradability to assure they are non-toxic, non-allergenic and do not induce adverse reactions in the body. The formation of new tissues should also match the degradation rate of the implanted materials to prevent further damage. Finally, good mechanical properties are also essential requirements of an ideal scaffold in bone tissue engineering [131–133].

Numerous biomaterials, such as metals, ceramics, polymers and composites, have been used in bone scaffolds. Metallic scaffolds are limited in use due to ion release, as well as limited bioactivity and biodegradation. Bioactive ceramics include calcium phosphates (CaP) and bioactive glasses, such as HA, beta-tricalcium phosphate (β-TCP), and a mixture of HA and β-TCP [131, 134, 135]. Polymers used in bone scaffolds include collagen, alginate, CS, fibrin, PCL, PEG, PVA, PPF, PLA, PGA and their copolymers [136]. The common biodegradable polymers used in bone scaffolds and their composites are discussed in this section.

Natural polymers can be used in bone tissue engineering due to their excellent biocompatibility, and they include agarose, collagen and CS [137]. Agarose, a type of polysaccharide, is extracted from seaweeds [137], and exhibits high turbidity and strong elasticity. It has been found that agarose confers toughness, ductility and a rubbery consistency for strains of ≤60% when incorporated into biphasic CaP and biphasic CaP–agarose scaffolds, which may be suitable for use in bone regeneration [138].

Collagen is a fibrous protein that is a major component in connective tissues. Collagen has been used widely for tissue-regeneration applications [137]. It is the most common protein in the body and provides strength and structural stability to tissues in the body including skin, blood vessels, tendon, cartilage and bone [139]. As a material for scaffolds in tissue engineering, collagen provides excellent biological characteristics because it improves cell attachment, growth and proliferation. Hence, collagen scaffolds have been used for many years in several *in vitro* and *in vivo* studies in skin regeneration, cartilage repair and many other tissues. While collagen possesses these advantageous characteristics it lacks the mechanical properties required for bone tissue engineering. Various approaches have been carried out to help strengthen collagen scaffolds, including crosslinking methods, yet the stiffness of crosslinked collagen scaffolds remains one order of magnitude lower than that for bone [140].

Next to cellulose, CS is one of the most abundant natural polymers. It has received considerable attention in bone tissue engineering due to its favourable antimicrobial properties and its degradation rate of 2–8 weeks (depending on the level of crosslinking), which closely resembles normal rates of bone repair [141, 142].

However, scarcity in bulk quantities and difficulties in the processability of natural polymers limit their use in tissue engineering. In addition, the degradation rate of natural polymer materials is dependent on the patient because the degradation of natural polymer materials is dependent upon enzymes, which vary from patient to patient. On the other hand, synthetic polymers exhibit predictable and reproducible physical, chemical and degradation properties that can be modified to meet the specific requirements of bone tissue recovery [139]. Synthetic polymers used in scaffold preparation include PGA, PLA and PCL [143].

PGA is the simplest linear aliphatic polyester. It has a highly crystalline structure, high melting point and low solubility in organic solvents. The degradation products of PGA are naturally resorbed by the body, which makes it a desirable biomaterial. As PGA degrades rapidly, between 4 months and 12 months, this characteristic can make it difficult to process and handle. PGA is however, commonly extruded as thin polymer strands ≈13 µm in diameter and moulded into non-woven mesh discs to produce a scaffold structure. This structure provides a high-porosity environment for cells to grow and proliferate. However, PGA can lack sufficient mechanical integrity. On the contrary, PLA–PGA copolymers have higher mechanical properties after going through two self-reinforcing methods: sintering and fibrillation. These methods induce highly oriented polymer morphology and consequently make this copolymer material more attractive for use in clinical applications. PLA–PGA copolymers are fabricated as non-woven mesh to engineer cartilage scaffolds and for appropriate construction. The interconnectivity, porosity, pore size and void space can be adjusted during the fabrication process [137].

PLA degrades slower than PGA due to its hydrophobic characteristic, which limits the water absorption of thin films and slows the hydrolysis rate of the polymer backbone. Based on available data, it can take anywhere from 12 months to >2 years for this material to degrade [137]. Although many researchers have tried to investigate porous PLA scaffolds for use in orthopaedic applications, PLA is employed primarily as a non-woven mesh for tissue engineering applications. *In vitro* cell culture experiments have depicted good osteoblast attachment and proliferation following the fabrication of continuous and uniform composite electrospun microfibres made of PLLA incorporated with HA nanopowder. It was also found that, after 7 days of culture, there was enhanced expression of ALP [144].

PCL is another type of synthetic polymer which, due to its biocompatibility, degradation (≤24 months) and mechanical strength, makes it a suitable and compatible material for orthopaedic applications. Due to its degradation rate it is used to copolymerise with other materials to have desired degradation properties. Honda and co-workers carried out a study using poly(L-lactic-ε-caprolactone) as a biodegradable sponge and implanted it into mice. After 4 weeks of the study, formation of cartilage-like structures within the polymer construct was observed [145].

These synthetic materials have shown considerable success because they can be fabricated with a tailored architecture and their degradation characteristics may be controlled by varying the polymer itself or the composition of the individual polymer [139]. For instance Canillas and co-workers investigated the use of the freeze–thaw method to produce PVA-ceramic scaffolds for bone tissue engineering [146]. This process led to an increase in the mechanical strength of the scaffold with an increasing number of freeze–thaw cycles without the need for chemical crosslinking agents, and may prove beneficial in bone repair due to the resultant breakdown characteristics. However, these synthetic materials have disadvantages, such as the risk of rejection due to reduced bioactivity. In addition, concerns exist about the degradation process of PLLA and PGA because they degrade by hydrolysis. As a result, CO_2 is formed and lowers the pH, which can result in cell and bone-tissue necrosis.

Generally, a single type of scaffold material does not satisfy the current requirements for an extracellular scaffold. Therefore, two or more materials are often used in the composition of a bone scaffold. Today, bone scaffolds are usually made of a composite of polymers and bioactive ceramics. The aim is to increase the mechanical stability of scaffolds and to improve tissue interaction [147–150].

CaP-polymer scaffolds take advantages of CaP and polymers to meet mechanical and physiological requirements of the host tissue. Polymers in CaP scaffolds increase the toughness and compressive strength similar to that of bone, whereas CaP can improve the bioactivity of the composites [131]. PLGA/β-TCP scaffolds were reported to improve ALP activity, and have great capacity to induce osteogenic differentiation [151]. PCL/β-TCP scaffolds were manufactured *via* a 3D printer and found to have good biocompatibility and suitable mechanical strength, and PCL/β-TCP scaffolds degraded faster than those made of pure PCL due to the addition of β-TCP [152]. PGA/β- TCP scaffolds were developed using solvent casting and particulate leaching methods, and the PGA/β-TCP scaffolds exhibited a strong ability for osteogenesis, mineralisation and biodegradation for bone replacement [153]. PPF-*co*-PCL/HA scaffolds were found to induce maturation of pre-osteoblasts and have the potential for use in bone tissue engineering [154]. PDLLA/β-TCP scaffolds were manufactured by selective-laser melting, and revealed good osteoinductive properties that promote complete bone healing [155]. PLA/HA scaffolds were fabricated by 3D printing, and revealed to have good osteogenic capability and biodegradation activity [156]. Similarly, HA/ polyurethane scaffolds [157] and HA/PCL [158] scaffolds proved beneficial for use in bone repair.

Biodegradable polymer-based bone scaffolds have demonstrated excellent formation of new bone. However, some challenges remain:

1. Biodegradable polymeric scaffolds show rapid strength degradation *in vivo* even if they have high initial strength [131].
2. Degradation of PLA and PGA creates a local acidic environment that can cause adverse tissue responses [11, 159].

Scaffolds made from polyesters can cause poor wetting, lack of cellular attachment and interaction due to their hydrophobicity.

2.6.2 Repair of peripheral nerves

There is a long history relating to injuries to peripheral nerves. Herophilus in 300 BC first traced the nerve to the spinal cord *via* meticulous dissection. Rhazes in 900 AD was the first to discover nerve repair. Cruikshank demonstrated the healing and recovery of the distal end of a neuron. In the early 1900s, Cajal introduced the concept of axon regeneration guided by chemotropic substances [160].

Peripheral-nerve injuries present issues both clinically and socioeconomically by affecting the patient's quality of life. These injuries can have devastating results on sensory and motor impairments, which affect 1 in 1,000 patients [161]. Unlike the central nervous system, the peripheral nervous system (PNS) can regenerate damaged tissues. However, this intrinsic ability to repair itself relates to various factors such as age, mechanism of injury and, most importantly, the proximity of the injury to the nerve cell body [162]. The Seddon classification is used to categorise the severity of such injuries: neuropraxia (grade 1) relates to radical-nerve compression resulting in blockade of fibre conduction; axonotmesis is grade 2; neurotmesis can be grade 3, 4 or 5 [163].

Following traumatic injury to the PNS, pathophysiological events occur in the injured nerve. This leads to Wallerian degeneration in the distal end of the nerve, followed by axon degeneration. Macrophages and monocytes migrate into nerve stumps, resulting in removal of the myelin sheath and axon debris. In addition, Schwann cells proliferate to form the bands of Büngner and produce neurotrophic factors and extracellular material to aid axon regeneration. For short gaps, the preferred methodology employs the use of end–end suturing. However, if the gap is large, suturing cannot be done because the transected nerve stumps cannot reconnect without exerting excessive tension. Defects >3 cm in length are treated by nerve autografts, which are the 'gold standard'. Unfortunately, nerve autografts inherit complications such as neuroma, secondary surgical procedures, donor-site morbidity, nerve-site mismatch and a limited amount of donor tissue [164]. Alternative approaches utilise allogenic grafts, which are isolated from cadavers and are of limited supply and which can cause immunological rejection. Fortunately, the concept of 'nerve tubes' has evolved to implement strategies to improve the regeneration process of the defective nerves. However, in the 1960s, the first nerve scaffolds were constructed from non-degradable polymers such as silicone, acrylic and PE. Despite limited nerve recovery, the non-degradable polymers required a second surgical procedure to remove the scaffold, and they also became detrimental to the patient by inducing neural toxicity and by restricting nerve remodelling [165]. Modern strategies are based on the utilisation of bioresorbable polymers, and several have been approved by the FDA. Designs are based on natural, synthetic or a combination of both polymeric materials such as type-1 collagen (Neuragen®, NeuroFlex™, Neuro-Matrix, NeuraWrap™, Neuromed®), porcine small-intestine submucosa (Surgis® Nerve cuff), PGA (Nerotube) and poly(D,L-lactide-co-ε-CL) (Neurolac®) [163, 166]. Various materials with regeneration capacity are listed in Table 2.4. Surgical procedures using such materials have already being done. For example, Ashley and co-workers used collagen Neurogen scaffolds to bridge the brachial plexus in five patients. They concluded that use of collagen tubes was simple and effective [167]. Hung and co-workers used a 4 × 20 mm bioresorbable Nerotube for the recovery of the median nerve in a 53-year-old female [168]. Additionally, Donoghue and co-workers utilised a bioresorbable Nerotube to bridge the median nerves of 43- and 61-year-old patients [169]. These nerve-bridging devices are 3D tubular constructs that provide improved mechanical stability, flexibility and guidance properties that allow repair of the peripheral nerve if direct repair by neurorrhaphy is not possible [170].

These devices have potential and share similar efficacy compared with autografts surgery, but they are limited to nerve gaps >20 mm due to the simplicity of the design. Current approaches involve developing nerve tubes by incorporating

Table 2.4: Capacity of biomaterials to regenerate *in vivo*.

Material	Species	Fabrication method	Size of defect	Regrowth rate	Reference
Poly-hydroxybutyrate	Cat	Sheets	Superficial radial nerve	12 months	[176]
Silk	Wistar rat	Solvent casting	8 mm sciatic nerve	4 weeks	[177]
CS	Rat	Extrusion	10 mm sciatic nerve	13 weeks	[178]
PLA	Rat	Melt-blow	7 mm facial nerve	13 weeks	[179]
PLLA	Rat	Extrusion	10 mm sciatic nerve	16 weeks	[180]
PLLA/PCL	Rat	Coaxial electrospinning	10 mm sciatic nerve	12 weeks	[181]

physical, biological, or chemical cues, or filling nerve tubes with collage and laminin gel, Schwann cells or growth factors. For example, Luo and co-workers constructed nerve conduits from hollow cellulose tubes comprising Schwann cells and pyrroloquinoline quinone (Pq) (an antioxidant that stimulates nerve growth factors). They applied the nerve-guiding conduit onto the sciatic nerves of rats and demonstrated that a conduit comprising nerve growth factor and Pq could repair and reconstruct nerve structure and function [177]. In addition, Xu and colleagues fabricated PLGA with nectin-like molecule 1 (NECL1), which is a member of the immunoglobulin superfamily and is responsible for cellular adhesion and communication. In addition, NECL1 is known to adhere to axons and Schwann cells. Xu and co-workers fabricated PLGA film with pre-coated NECL1 to mimic the physiological environment of natural axons *in vivo*. They showed that PLGA/NECL1 conduits improved Schwann-cell aggregation and recovery of the nerve in comparison with PLGA [178]. Schmidt and co-workers utilised PLGA films and electrically conductive polypyrrole to enhance neurite length on PC12 cells *in vivo*. They showed an increase in neurite length in comparison with polystyrene controls [179]. Catrina and co-workers engineered collagen and silk fibroin scaffolds loaded with neurotrophic factors and glial cell line-derived neurotrophic factors (GDNF) to induce axonal repair and regeneration. These scaffolds were constructed to provide sustained release to achieve the discrete kinetics of neurotrophic factors and GDNF [180].

Constructing a conduit possessing intraluminal cues is only one of several strategies that must be considered to generate a conduit permitting axon growth. Semipermeable materials *via* inclusion of pores are an essential characteristic incorporated within the conduit. Not only do they promote cell attachment and axon regeneration, they also aid the diffusion of nutrients and waste products. It has been found that pore size also plays a vital part most notably in natural materials such as collagen [181, 182].

Other modifications involve using laminin or fibronectin material for coatings to induce nerve regeneration as well as cell attachment and migration. Also, the microarchitecture can be modified by incorporating filaments, sponges and multichannel nerve tube structures [183]. This modification was designed to induce regeneration of nerves across lesions, and to direct the 'sprouting of axons' from the proximal nerve end. The desired effect of these nerve-guidance conduits is to increase the recovery speed and length of axon regeneration. As mentioned above, various parameters can be modified to improve nerve regeneration among nerve conduits, including porosity, coatings, microfabrication, intraluminal cues and fibres. Introducing fibre bundles into the conduit significantly improves the longitudinal cellular migration of neurons and Schwann cells. Chwalek and co-workers analysed the impact of fibre profiling on cellular growth on PLA fibres *in vitro*. They concluded that neuron outgrowth and the morphological design of PLA fibres dictate nerve regeneration [184]. The aim of constructing synthetic scaffolds is to stimulate the morphology of native nerves and accelerate repair of damaged nerves. Numerous strategies and an abundance of literature have been explored. However, these improvements are limited to simple circular geometries. This could be attributed to limitations inherited with current material-processing techniques. Fortunately, there is ongoing research of utilising fibre photonic apparatuses. For instance, Koppes and co-workers employed thermal drawing processing (TDP) to produce various cylindrical and rectangular core neural scaffolds. They concluded that the ability to direct integration of the micro-grooved topography of the scaffold allowed for improved Schwann-cell migration [185]. In addition, 3D printing technology has been found to be a promising method to produce complex biomedical devices. In particular, PLA 3D printing has been used for the improvement of implants, scaffolds and drug-delivery systems but unfortunately complex tissue such as bone, cartilage and nerves has proved to be challenging due to technological limitations [186]. Artificial PLA conduits have been explored extensively in peripheral-nerve regeneration due to their flexibility, mechanical characteristics and lack of antigenicity. *In vivo* studies have been investigated, but Tyler and co-workers stated that gap lengths >10 mm in a model of sciatic nerves resulted in consistent failure [187]. This problem emphasises the importance of optimising the inner architecture of the conduit to induce nerve regeneration. Cai and co-workers utilised PLA conduits to induce axonal migration in a rat model [188]. Similar strategies have also being employed on spinal-cord injuries. Oudega and co-workers used resorbable PLA_{50}, $PLA_{100/10}$ and PDLA tubular scaffolds with implanted Schwann cells into transected thoracic spinal cords in rats for ≤4 months and demonstrated increased axonal regeneration [189]. In addition, Goulart and co-workers transected the left sciatic nerve of a C57BL/6 mouse. They demonstrated that the PLA conduit could not only be fabricated in a controlled environment and was a suitable substrate for cell survival, but also induced axonal growth [190]. Studies undertaken by Lu and co-workers utilised a PLA conduit fabricated by 'microbraiding'. This improved the mechanical strength of the polymer and, as a result of *in vivo* studies, demonstrated improved regeneration at 8 weeks after surgery [191]. In addition, PLA was used to bridge a 20-mm nerve-gap defect. The conduits contained a macrospore exterior and interconnected microspores on the interior to provide a sufficient outflow rate and then a sufficient inflow rate. This conduit was fabricated *via* immersion precipitation and had 80% functional recovery after 18 months [192, 193]. Another important parameter to take into consideration is the swelling and degradation of a nerve tube. If there is rapid degradation, this may be attributed to swelling and if the degradation is too slow it may lead to compression or a chronic foreign-body reaction. To avoid this scenario, the swelling and degradation properties may be optimised *via* tube dimension and copolymer ratio. The ideal nerve tube should remain intact until axons fully regenerate across the nerve gap [183, 194, 195].

2.7 Summary

Biodegradable implants can provide support, deliver pharmacological agents, and degrade after service, offering a promising alternative for patients. However, some challenges remain. Cost is one of the major issues. This results from the high cost of materials and manufacture, and the additional, or sometimes more time-consuming, procedures required to surgically implant biodegradable implants. It is therefore important to lower these costs to spread the use of biodegradable implants. Another reported problem related to biodegradable implants is complications due to delayed or incomplete degradation processes. Ideally, biodegradable implants should degrade at rates that satisfy the conditions of healing and load-carrying ability. However, incomplete degraded device parts have been reported to remain within the body and cause an inflammatory response that interferes with recovery of the injured site. The complex degradation mechanism of polymer composites, which has often been adopted by biodegradable implants, requires thorough study. Moreover, the mechanical strength of biodegradable implants, especially polymer implants, needs to be improved. So far, the use of biodegradable polymeric implants has been limited due to the weakness of biodegradable polymers. Reinforced biodegradable polymers make it possible for high-load applications such as long-bone fixation. Finally, fabrication methods need to be developed or optimised to suit the biodegradable polymers used.

References

1.   E. Regar, G. Sianos and P.W. Serruys, *British Medical Bulletine*, 2001, **59**, 227.
2.   W. Khan, M. Kapoor and N. Kumar, *Acta Biomaterialia*, 2007, **3**, 4, 541.
3.   C.M. Agrawal, *The Journal of The Minerals, Metals & Materials Society*, 1998, **50**, 1, 31.
4.   D. Bombač, M. Brojan, P. Fajfar, F. Kosel and R. Turk, *RMZ – Materials and Geoenvironment*, 2007, **54**, 54, 471.
5.   A. Metzger, *Biomedical Engineering*, 1976, **11**, 301.

6. W. Khan, E. Muntimadugu, M. Jaffe and J.A. Domb in *Focal Controlled Drug Delivery, Advances in Delivery Science and Technology*, Eds., A.J. Domb and W. Khan, Springer, New York, NY, USA, 2014, p.33.
7. S. Lyu and D. Untereker, *International Journal of Molecular Sciences*, 2009, **10**, 9, 4033.
8. C.C. Chu in *Wound Closure Biomaterials and Devices*, Eds., C-C. Chu and J.A. Von, CRC Press, Boca Raton, FL, USA, 1997, p.1.
9. K. Tuzlakoglu and R.L. Reis in *Biodegradable Systems in Tissue Engineering and Regenerative Medicine*, Eds., R.L. Reis and J.S. Roman, CRC Press, Boca Raton, FL, USA, 2005, p.198.
10. C.K.S. Pillai and C.P. Sharma, *Journal of Biomaterials Applications*, 2010, **25**, 4, 291.
11. Y. Chen, L.M. Geever, J.A. Killion, J.G. Lyons, C.L. Higginbotham and D.M. Devine, *Polymer–Plastics Technology and Engineering*, 2016, **55**, 10, 1057.
12. S. Suzuki and Y. Ikada in *Biomaterials for Surgical Operation*, Eds., S. Suzuki and Y. Ikada, Humana Press, Totowa, NJ, USA, 2012, p.189.
13. J. Hochberg, K.M. Meyer and M.D. Marion, *Surgical Clinics of North America*, 2009, **89**, 3, 627.
14. K.K. Outlaw, A.R. Vela and J.P. O'Leary, *The American Surgeon*, 1998, **64**, 4, 348.
15. E.S. Debus, D. Geiger, M. Sailer, J. Ederer and A. Thiede, *European Surgical Research*, 1997, **29**, 1, 52.
16. F.B. Niessen, P.H. Spauwen and M. Kon, *Annals of Plastic Surgery*, 1997, **39**, 3, 254.
17. J. Kangas, S. Paasimaa, P. Mäkelä, J. Leppilahti, P. Törmälä, T. Waris and N. Ashammakhi, *Journal of Biomedical Materials Research*, 2001, **58**, 1, 121.
18. P. Mäkelä, T. Pohjonen, P. Törmälä, T. Waris and N. Ashammakhi, *Biomaterials*, 2002, **23**, 12, 2587.
19. Y. Harada, K. Kadono, T. Terao, M. Suzuki, Y. Ikada and N. Tomita, *Journal of Veterinary Medical Science*, 2013, **75**, 9, 1187.
20. A. Obermeier, J. Schneider, S. Wehner, F.D. Matl, M. Schieker, R. von Eisenhart-Rothe, A. Stemberger and R. Burgkart, *PLoS ONE*, 2014, **9**, 7, e101426.
21. A.M. Behrens, B.J. Casey, M.J. Sikorski, K.L. Wu, W. Tutak, A.D. Sandler and P. Kofinas, *ACS Macro Letters*, 2014, **3**, 3, 249.
22. C. Jaboro, *Polymer Solutions Incorporated Newsletter*, 2014, 30th March 2014.
23. Y. Zhang, S. Liu, X. Wang, Z. Zhang, X. Jing, P. Zhang and Z. Xie, *Chinese Journal of Polymer Science*, 2014, **32**, 8, 1111.
24. J. Zhang, X. Wang, T. Liu, S. Liu and X. Jing, *Drug Delivery*, 2016, **23**, 3, 794.
25. H. Neumann, A.P. Schulz, J. Gille, M. Klinger, C. Jürgens, N. Reimers and B. Kienast, *Bone & Joint Research*, 2013, **2**, 2, 26.
26. C.A. Campbell and K.Y. Lin, *Craniomaxillofacial Trauma & Reconstruction*, 2009, **2**, 1, 41.
27. G. Becker, A. Calvis, L. Hazlett, M. Verzi and M. Paliwal in *Proceedings of the 40th Annual Northeast Bioengineering Conference (NEBEC)*, 25–27th April, Northeastern University's College of Engineering, Boston, MA, USA, 2014, p.1.
28. S.J. Ferguson, U.P. Wyss and D.R. Pichora, *Medical Engineering & Physics*, 1996, **18**, 3, 241.
29. J.C. Middleton and A.J. Tipton, *Biomaterials*, 2000, **21**, 23, 2335.
30. J.S. Hayes and R.G. Richards, *Expert Review of Medical Devices*, 2010, **7**, 1, 131.
31. J.S. Hayes and R.G. Richards, *Expert Review of Medical Devices*, 2010, **7**, 6, 843.
32. J.S. Hayes, U. Seidenglanz, A.I. Pearce, S.G. Pearce, C.W. Archer and R.G. Richards, *European Cells and Materials*, 2010, **19**, 0, 117.
33. J.S. Hayes, C.W. Archer and R.G. Richards, *European Cells and Materials*, 2007, **13**, Supplement 2, 67.
34. J.S. Hayes, D.I. Vos, J. Hahn, S.G. Pearce and R.G. Richards, *European Cells and Materials*, 2009, **18**, 0, 15.

35. C.P. Case, V.G. Langkamer, C. James, M.R. Palmer, A.J. Kemp, P.F. Heap and L. Solomon, *Bone & Joint Journal*, 1994, **76**, 5, 701.
36. D.S. Jorgenson, M.H. Mayer, R.G. Ellenbogen, J.A. Centeno, F.B. Johnson, F.G. Mullick and P.N. Manson, *Plastic and Reconstructive Surgery*, 1997, **99**, 4, 976.
37. D.I. Vos and M.H.J. Verhofstad, *European Journal of Trauma and Emergency Surgery*, 2013, **39**, 4, 327.
38. D. Maier, K. Izadpanah, P. Ogon, M. Mützel, J. Bayer, N.P. Südkamp and M. Jaeger, *Archives of Orthopaedic and Trauma Surgery*, 2015, **135**, 7, 953.
39. S. Pina and J. Ferreira, *Journal of Healthcare Engineering*, 2012, **3**, 2, 243.
40. J.H. Clark and J.J.E. Hardy in *Sustainable Development in Practice: Case Studies for Engineers and Scientists*, Eds., A. Azapagic, S. Perdan and R. Clift, John Wiley & Sons, Hoboken, NJ, USA, 2004, p.250.
41. H.E. Aryan, D.C. Lu, F.L. Acosta, Jr., R. Hartl, P.W. McCormick and C.P. Ames, *Spine*, 2007, **32**, 10, 1084.
42. G.R. Zaveri and M. Ford, *Journal of Spinal Disorders & Techniques*, 2001, **14**, 1, 10.
43. J.C.H. Goh, H. Wong, A. Thambyah and C-S. Yu, *Spine*, 2000, **25**, 1, 35.
44. S. Dennis, R. Watkins, S. Landaker, W. Dillin and D. Springer, *Spine*, 1989, **14**, 8, 876.
45. S. Kumar Tripathi, S. Narayan Nanda, S. Vatchha, A. Kohli, P. Pradhan, S. Muzammil Shiraz and C. Author, *The Journal of Spinal Surgery*, 2014, **1**, 4, 154.
46. B. Brożyna, H. Szymańska, K. Ptaszyński, M. Woszczyński, J. Lechowska-Piskorowska, M. Gajewska, J. Rostkowska, K. Chełmiński, W. Bulski and R. Krajewski, *Oral Surgery, Oral Medicine, Oral Pathology and Oral Radiology*, 2015, **120**, 4, 443.
47. E. Ellis and L.R. Walker, *Journal of Oral and Maxillofacial Surgery*, 1996, **54**, 7, 864.
48. G.D. Wood, *British Journal of Oral and Maxillofacial Surgery*, 2006, **44**, 1, 38.
49. R.K. Bali, P. Sharma, S. Jindal and S. Gaba, *National Journal of Maxillofacial Surgery*, 2013, **4**, 2, 167.
50. S.K. Elhalawany, B. Tarakji, S.N. Azzeghaiby, I. Alzoghaibi, K. Baroudi and M.Z. Nassani, *Nigerian Medical Journal*, 2015, **56**, 1, 48.
51. M. Bayat, A. Garajei, K. Ghorbani and M.H.K. Motamedi, *Journal of Oral and Maxillofacial Surgery*, 2010, **68**, 7, 1573.
52. K. Savage, Z.M. Sardar, T. Pohjonen, G.S. Sidhu, B.D. Eachus and A. Vaccaro, *Clinical Spine Surgery*, 2014, **27**, 2, E66.
53. J.Y. Rho, R.B. Ashman and C.H. Turner, *Journal of Biomechanics*, 1993, **26**, 2, 111.
54. Y. Matsusue, T. Nakamura, S. Suzuki and R. Iwasaki, *Clinical Orthopaedics and Related Research*, 1996, **322**, 166.
55. T. Ma'ruf, W. Siswomihardjo, M.H. Soesatyo and A.E. Tontowi in *Proceedings of the 3rd International Conference on Instrumentation, Communications, Information Technology and Biomedical Engineering (ICICI-BME)*, ITB Bandung, Indonesia, 2013, p.246.
56. J.F. Mano, R.A. Sousa, L.F. Boesel, N.M. Neves and R.L. Reis, *Composites Science and Technology*, 2004, **64**, 789.
57. Y.H. An, S.K. Woolf and R.J. Friedman, *Biomaterials*, 2000, **21**, 2635.
58. G. Perego and G.D. Cella in *Poly(Lactic Acid): Synthesis, Structures, Properties, Processing, and Applications*, Eds., R. Auras, L.T. Lim, S.E.M. Selke and H. Tsuji, John Wiley & Sons, NJ, USA, 2010, p.141.
59. E. Aydin, J.A. Planell and V. Hasirci, *Journal of Materials Science: Materials in Medicine*, 2011, **22**, 11, 2413.
60. T. Czigány, J.G. Kovács and T. Tábi in *Proceedings of the 19th International Conference on Composite Materials*, Montreal, Canada, 2013, p.4377.

61. R.M. Felfel, I. Ahmed, A.J. Parsons, G.S. Walker and C.D. Rudd, *Journal of the Mechanical Behavior of Biomedical Materials*, 2011, **4**, 7, 1462.
62. S. Sukegawa, T. Kanno, N. Katase, A. Shibata, Y. Takahashi and Y. Furuki, *The Journal of Craniofacial Surgery*, 2016, **27**, 6, 1391.
63. W.J. Ciccone, C. Motz, C. Bentley and J.P. Tasto, *Journal of the American Academy of Orthopaedic Surgeons*, 2001, **9**, 280.
64. A.R. Vaccaro, M.M. Robbins, L. Madigan, T.J. Albert, W. Smith and A.S. Hilibrand, *Neurosurgical Focus*, 2004, **16**, 3, E7.
65. A.R. Vaccaro, K. Singh, R. Haid, S. Kitchel, P. Wuisman, W. Taylor, C. Branch and S. Garfin, *Spine Journal*, 2003, **3**, 3, 227.
66. H.H. Peltoniemi, D. Hallikainen, T. Toivonen, P. Helevirta and T. Waris, *Journal of Cranio-Maxillofacial Surgery*, 1999, **27**, 1, 42.
67. A. Xue, J. Koshy, J. Weathers, E. Wolfswinkel, Y. Kaufman, S. Sharabi, R. Brown, M. Hicks and L. Hollier, *Craniomaxillofacial Trauma and Reconstruction*, 2014, **7**, 1, 27.
68. M. Bao, X. Wang, H. Yuan, X. Lou, Q. Zhao and Y. Zhang, *Journal of Materials Chemistry B*, 2016, **4**, 31, 5308.
69. Y. Zhang, J. Wang, J. Wang, X. Niu, J. Liu, L. Gao, X. Zhai and K. Chu, *Cell and Tissue Banking*, 2015, **16**, 4, 615.
70. R. Laughlin, M. Block and R. Wilk, *Journal of Oral and Maxillofacial Surgery*, 2007, **56**, 89.
71. E. Nkenke, E. Vairaktaris, C. Knipfer, F. Stelzle, S. Schwarz, I. Eyüpoglu, O. Ganslandt and T. Leis, *Neurocirugia*, 2011, **22**, 6, 498.
72. M.J. Imola, D.D. Hamlar, W. Shao, K. Chowdhury and S. Tatum, *Archives of Facial Plastic Surgery*, 2001, **3**, 2, 79.
73. J.M. Pensler, *Journal of Craniofacial Surgery*, 1997, **8**, 2, 129.
74. O.M. Böstman and H.K. Pihlajamäki, *Clinical Orthopaedics and Related Research*, 2000, **371**, 216.
75. D.R. Holmes, R.E. Vlietstra, H.C. Smith, G.W. Vetrovec, K.M. Kent, M.J. Cowley, D.P. Faxon, A.R. Gruentzig, S.F. Kelsey and K.M. Detre, *The American Journal of Cardiology*, 1984, **53**, 12, 77C.
76. C.M. Campos, T. Muramatsu, J. Iqbal, Y-J. Zhang, Y. Onuma, H.M. Garcia-Garcia, M. Haude, P.A. Lemos, B. Warnack and P.W. Serruys, *International Journal of Molecular Sciences*, 2013, **14**, 12, 24492.
77. V. Farooq, B.D. Gogas and P.W. Serruys, *Circulation: Cardiovascular Interventions*, 2011, **4**, 2, 195.
78. B.R. Brodie, Y. Pokharel, A. Garg, G. Kissling, C. Hansen, S. Milks, M. Cooper, C. McAlhany and T.D. Stuckey, *Journal of Interventional Cardiology*, 2014, **27**, 1, 21.
79. K. Yamaji, T. Kimura, T. Morimoto, Y. Nakagawa, K. Inoue, S. Kuramitsu, Y. Soga, T. Arita, S. Shirai, K. Ando, K. Kondo, K. Sakai, M. Iwabuchi, H. Yokoi, H. Nosaka and M. Nobuyoshi, *Journal of the American Heart Association*, 2012, **1**, 5, e004085.
80. P. Smits, G.J. Vlachojannis, E.P. McFadden, K.-J. Royaards, J. Wassing, K.S. Joesoef, C. Van Mieghem and M. Van De Ent, *JACC: Cardiovascular Interventions*, 2015, **8**, 9, 1157.
81. Y. Onuma and P.W. Serruys in *Coronary Stenosis Imaging, Structure and Physiology*, Eds., J. Escaned and P.W. Serruys, Europa Digital Publishing, Toulouse, France, 2015.
82. W.J. van der Giessen, A.M. Lincoff, R.S. Schwartz, H.M. van Beusekom, P.W. Serruys, D.R. Holmes, S.G. Ellis and E.J. Topol, *Circulation*, 1996, **94**, 7, 1690.
83. A.M. Lincoff, J.G. Furst, S.G. Ellis, R.J. Tuch and E.J. Topol, *Journal of the American College of Cardiology*, 1997, **29**, 4, 808.
84. J.M. Raquez, Y. Habibi, M. Murariu and P. Dubois, *Progress in Polymer Science*, 2013, **38**, 10–11, 1504.
85. A. Abizaid and J.R. Costa, *Circulation: Cardiovascular Interventions*, 2010, **3**, 4, 384.
86. J.A. Ormiston and P.W.S. Serruys, *Circulation: Cardiovascular Interventions*, 2009, **2**, 255.

87. A.M. Sammel, D. Chen and N. Jepson, *Heart Lung and Circulation*, 2013, **22**, 7, 495.
88. R.D. Alexy and D.S. Levi, *BioMed Research International*, 2013, Article ID:137985.
89. P. Gasior, Y. Cheng, L. Wang, G. Feng, G.B. Conditt, J. McGregor, F.D. Kolodgie, R. Virmani, J. Granada and G.L. Kaluza, *Journal of the American College of Cardiology*, 2015, **66**, 15, B58.
90. Y. Zhang, C. V Bourantas, V. Farooq, T. Muramatsu, R. Diletti, Y. Onuma, H.M. Garcia-Garcia and P.W. Serruys, *Medical Devices*, 2013, **6**, 37.
91. B. O'Brien and W. Carroll, *Acta Biomaterialia*, 2009, **5**, 4, 945.
92. S. Garg and P. Serruys, *Minerva Cardioangiologica*, 2009, **57**, 5, 537.
93. M. O'Riordan in *The Bioresorbable Stent Story So Far: What Promise? What Price? TCTMD*, 2016, 17th August 2016. https://www.tctmd.com/news/bioresorbable-stent-story-so-far-what-promise-what-price.
94. Y. Chen, L.M. Geever, J.A. Killion, J.G. Lyons, C.L. Higginbotham and D.M. Devine, *Polymer Composites*, 2015, DOI:10.1002/pc.23794.
95. B. Stepak, A.J. Antończak, M. Bartkowiak-Jowsa, J. Filipiak, C. Pezowicz and K.M. Abramski, *Archives of Civil and Mechanical Engineering*, 2014, **14**, 2, 317.
96. B. Heublein, R. Rohde, V. Kaese, M. Niemeyer, W. Hartung and A. Haverich, *Heart (British Cardiac Society)*, 2003, **89**, 6, 651.
97. R. Barenghi, S. Beke, I. Romano, P. Gavazzo, B. Farkas, M. Vassalli, F. Brandi and S. Scaglione, *BioMed Research Internationalternational*, 2014, **2014**, 9.
98. I. Valverde, G. Gomez, J.F. Coserria, C. Suarez-Mejias, S. Uribe, J. Sotelo, M.N. Velasco, J.Santos De Soto, A-R. Hosseinpour and T. Gomez-Cia, *Catheterization and Cardiovascular Interventions*, 2015, **85**, 6, 1006.
99. G. Li, Y. Li, P. Lan, J. Li, Z. Zhao, X. He, J. Zhang and H. Hu, *Journal of Biomedical Materials Research, Part A*, 2014, **102**, 4, 982.
100. F. Schreiber, P. Schuster, M. Borinski, F. Vogt, R. Blindt and T. Gries, *Autex Research Journal*, 2010, **10**, 3, 73.
101. Y. Onuma and P.W. Serruys, *Circulation*, 2011, **123**, 779.
102. E. Spuentrup, A. Ruebben, A. Mahnken, M. Stuber, C. Kölker, T.H. Nguyen, R.W. Günther and A. Buecker, *Circulation*, 2005, **111**, 8, 1019.
103. M. Niaounakis in *Biopolymers: Applications and Trends*, Ed., M. Niaounakis, Elsevier, Amsterdam, The Netherlands, 2015, p.291.
104. J.A. Ormiston, B. Webber and M.W.I. Webster, *JACC: Cardiovascular Interventions*, 2011, **4**, 12, 1310.
105. M. Vert, *European Polymer Journal*, 2015, **68**, 516.
106. R.W. Van Holten and J.C. Patel, inventors; Ethicon Inc., assignee; US9238088B2, 2016.
107. E. Lih, S.H. Oh, Y.K. Joung, J.H. Lee and D.K. Han, *Progress in Polymer Science*, 2015, **44**, 28.
108. Y.S. Chung, S-N. Park, J.H. Ko, S.H. Bae, S. Lee, I.K. Shim and S.C. Kim, *Journal of Surgical Research*, 2016, **205**, 2, 341.
109. D. Niwa, M. Koide, T. Fujie, N. Goda and S. Takeoka, *Journal of Biomedical Materials Research, Part B: Applied Biomaterials*, 2013, **101**, 7, 1251.
110. T. Komachi, H. Sumiyoshi, Y. Inagaki, S. Takeoka, Y. Nagase and Y. Okamura, *Journal of Biomedical Materials Research, Part B: Applied Biomaterials*, 2016, DOI:10.1002/jbm.b.33714.
111. M. Kessler, E. Esser, J. Groll and J. Tessmar, *Journal of Biomedical Materials Research, Part B: Applied Biomaterials*, 2015, **104B**, 1563.
112. A. Hinoki, A. Saito, M. Kinoshita, J. Yamamoto, D. Saitoh and S. Takeoka, *British Journal of Surgery*, 2016, **103**, 6, 692.
113. Z. Chen, J. Zheng, J. Zhang and S. Li, *Interactive Cardiovascular and Thoracic Surgery*, 2015, **21**, 5, 565.
114. C. Ni, R. Lu, L. Tao, G. Shi, X. Li and C. Qin, *Polymer Bulletin*, 2015, **72**, 6, 1515.

115. Z. Song, B. Shi, J. Ding, X. Zhuang, X. Zhang, C. Fu and X. Chen, *Chinese Journal of Polymer Science*, 2015, **33**, 4, 587.
116. S.T. Jayanth, A. Pulimood, D. Abraham, A. Rajaram, M.J. Paul and A. Nair, *Annals of Medicine and Surgery*, 2015, **4**, 4, 388.
117. J.E. Ko, Y-G. Ko, W. Il Kim, O.K. Kwon and O.H. Kwon, *Journal of Biomedical Materials Research, Part B: Applied Biomaterials*, 2016, DOI:10.1002/jbm.b.33726.
118. D. Kim, S. Bang, C.J. Kim, W. Il Kim and O.H. Kwon, *Textile Science and Engineering*, 2015, **52**, 2, 104.
119. J.D. Cabral, M.A. McConnell, C. Fitzpatrick, S. Mros, G. Williams, P.J. Wormald, S.C. Moratti and L.R. Hanton, *Journal of Biomedical Materials Research, Part A*, 2015, **103**, 8, 2611.
120. R. Shi, J. Xue, H. Wang, R. Wang, M. Gong, D. Chen, L. Zhang and W. Tian, *Journal of Materials Chemistry B*, 2015, **3**, 19, 4063.
121. C-H. Chen, S-H. Chen, K.T. Shalumon and J-P. Chen, *Colloids and Surfaces B: Biointerfaces*, 2015, **133**, 221.
122. B. Yang, C. Gong, Z. Qian, X. Zhao, Z. Li, X. Qi, S. Zhou, Q. Zhong, F. Luo and Y. Wei, *BMC Biotechnology*, 2010, **10**, 1, 65.
123. N. Bölgen, I. Vargel, P. Korkusuz, Y.Z. Menceloğlu and E. Pişkin, *Journal of Biomedical Materials Research, Part B: Applied Biomaterials*, 2007, **81**, 2, 530.
124. R. Langer and J.P. Vacanti, *Science*, 1993, **260**, 920.
125. I. Armentano, M. Dottori, E. Fortunati, S. Mattioli and J.M. Kenny, *Polymer Degradation and Stability*, 2010, **95**, 11, 2126.
126. L. Xiaoming, C. Rongrong, S. Lianwen and E. Katerina, *International Journal of Polymer Science*, 2014, **2014**, 829145, 1.
127. M. Santoro and G. Perale in *Durability and Reliability of Medical Polymers*, Woodhead Publishing, Cambridge, UK, 2012, p.119.
128. C. Sala, M. Ribes, T. Muiños, L. Sancho and P. Chicón, *Journal of Biochips & Tissue Chips*, 2013, **2013**, S2.
129. F. Ghorbani, H. Nojehdehian and A. Zamanian, *Materials Science and Engineering: C*, 2016, **69**, 208.
130. A.R. Santos in *Tissue Engineering*, Ed., D. Enerli, InTech, Rijeka, Croatia, 2010, p.225.
131. S. Bose, M. Roy and A. Bandyopadhyay, *Trends in Biotechnology*, 2012, **30**, 10, 546.
132. N. Saranya, S. Saravanan, A. Moorthi, B. Ramyakrishna and N. Selvamurugan, *Journal of Biomedical Nanotechnology*, 2011, **7**, 2, 238.
133. H. Liu, L. Zhang, P. Shi, Q. Zou, Y. Zuo and Y. Li, *Journal of Biomedical Materials Research, Part B: Applied Biomaterials*, 2010, **95B**, 1, 36.
134. I. Denry and L.T. Kuhn, *Dental Materials*, 2016, **32**, 1, 43.
135. S.V. Dorozhkin, *Biomaterials*, 2010, **31**, 7, 1465.
136. S-H. Lee and H. Shin, *Advanced Drug Delivery Reviews*, 2007, **59**, 4–5, 339.
137. H-Y. Cheung and T-P. Lu, *Composites Part B: Engineering*, 2007, **38**, 3, 291.
138. J.A. Puértolas, J.L. Vadillo, S. Sánchez-Salcedo, A. Nieto, E. Gómez-Barrena and M. Vallet-Regí, *Acta Biomaterialia*, 2011, **7**, 2, 841.
139. F.J. O'Brien, *Materials Today*, 2011, **14**, 3, 88.
140. A.A. Al-Munajjed and F.J. O'Brien, *Journal of the Mechanical Behavior of Biomedical Materials*, 2009, **2**, 2, 138.
141. B. McKibbin, *The Journal of Bone and Joint Surgery*, 1978, **60B**, 2, 150.
142. A.K. Azab, B. Orkin, V. Doviner, A. Nissan, M. Klein, M. Srebnik and A. Rubinstein, *Journal of Controlled Release*, 2006, **111**, 3, 281.
143. M. Sokolsky-Papkov, K. Agashi, A. Olaye, K. Shakesheff and A.J. Domb, *Advanced Drug Delivery Reviews*, 2007, **59**, 4–5, 187.

144. H-W. Kim, H-H. Lee and J.C. Knowles, *Journal of Biomedical Materials Research, Part A*, 2006, **79A**, 3, 643.
145. M. Honda, T. Yada, M. Ueda and K. Kimata, *Journal of Oral and Maxillofacial Surgery*, 2000, **58**, 7, 767.
146. M. Canillas, G.G. De Lima, M.A. Rodriguez, M.J.D. Nugent and D.M. Devine, *Journal of Polymer Science, Part B: Polymer Physics*, 2016, **54**, 7, 761.
147. H-W. Kim, J.C. Knowles and H-E. Kim, *Biomaterials*, 2004, **25**, 7, 1279.
148. A.R. Boccaccini, J.J. Blaker, V. Maquet, R.M. Day and R. Jérôme, *Materials Science and Engineering: C*, 2005, **25**, 1, 23.
149. W. Xue, A. Bandyopadhyay and S. Bose, *Journal of Biomedical Materials Research, Part B: Applied Biomaterials*, 2009, **91**, 2, 831.
150. M.W. Laschke, A. Strohe, M.D. Menger, M. Alini and D. Eglin, *Acta Biomaterialia*, 2010, **6**, 6, 2020.
151. H. Wu, G.H. Liu, Q. Wu and B. Yu, *Genetics and Molecular Research*, 2014, **14**, 4, 11933.
152. L. Lu, Q. Zhang, D. Wootton, R. Chiou, D. Li, B. Lu, P. Lelkes and J. Zhou, *Journal of Materials Science: Materials in Medicine*, 2012, **23**, 9, 2217.
153. H. Cao and N. Kuboyama, *Bone*, 2010, **46**, 2, 386.
154. J. Becker, L. Lu, M.B. Runge, H. Zeng, M.J. Yaszemski and M. Dadsetan, *Journal of Biomedical Materials Research, Part A*, 2015, **103**, 8, 2549.
155. R. Smeets, M. Barbeck, H. Hanken, H. Fischer, M. Lindner, M. Heiland, M. Wöltje, S. Ghanaati and A. Kolk, *Journal of Biomedical Materials Research, Part B: Applied Biomaterials*, 2016, DOI:10.1002/jbm.b.33660.
156. H. Zhang, X. Mao, Z. Du, W. Jiang, X. Han, D. Zhao, D. Han and Q. Li, *Science and Technology of Advanced Materials*, 2016, **17**, 1, 136.
157. A. Asefnejad, A. Behnamghader, M.T. Khorasani and B. Farsadzadeh, *International Journal of Nanomedicine*, 2011, **6**, 93.
158. M.I. Hassan, T. Sun, N. Sultana, M.I. Hassan, T. Sun and N. Sultana, *Journal of Nanomaterials*, 2014, **2014**, 1.
159. S.R. Regueros, M. Albersen, S. Manodoro, S. Zia, N.I. Osman, A.J. Bullock, C.R. Chapple, J. Deprest and S. MacNeil, *BioMed Research International*, 2014, **2014**, 853610.
160. P. Mailänder and S. Zimmermann, *Zentralblatt für Chirurgie*, 2005, **130**, 1, W2.
161. C.J. Pateman, A.J. Harding, A. Glen, C.S. Taylor, C.R. Christmas, P.P. Robinson, S. Rimmer, F.M. Boissonade, F. Claeyssens and J.W. Haycock, *Biomaterials*, 2015, **49**, 77.
162. A. Faroni, S.A. Mobasseri, P.J. Kingham and A.J. Reid, *Advanced Drug Delivery Reviews*, 2015, **82**, 160.
163. V. Chiono and C. Tonda-Turo, *Progress in Neurobiology*, 2015, **131**, 87.
164. A.R. Nectow, K.G. Marra and D.L. Kaplan, *Tissue Engineering, Part B: Reviews*, 2012, **18**, 1, 40.
165. X. Gu, F. Ding, Y. Yang and J. Liu in *Neural Regeneration*, Eds., K-F. So and X-M. Xu, Elsevier, Amsterdam, The Netherlands, 2015, p.73.
166. S. Kehoe, X.F. Zhang and D. Boyd, *Injury*, 2012, **43**, 5, 553.
167. W.W. Ashley, T. Weatherly and T.S. Park, *Journal of Neurosurgery*, 2006, **105**, 6, 452.
168. V. Hung and A.L. Dellon, *Journal of Hand Surgery*, 2008, **33**, 3, 313.
169. N. Donoghoe, G. Rosson and L. Dellon, *Microsurgery*, 2007, **27**, 595.
170. M.F. Maitz, *Biosurface and Biotribology*, 2015, **1**, 3, 161.
171. A. Hazari, G. Johansson-Rudén, K. Junemo-Bostrom, C. Ljungberg, G. Terenghi, C. Green and M. Wiberg, *Journal of Hand Surgery*, 1999, **24B**, 3, 291.
172. W. Huang, R. Begum, T. Barber, V. Ibba, N.C.H. Tee, M. Hussain, M. Arastoo, Q. Yang, L.G. Robson, S. Lesage, T. Gheysens, N.J.V. Skaer, D.P. Knight and J.V. Priestley, *Biomaterials*, 2012, **33**, 1, 59.

173. K. Haastert-Talini, S. Geuna, L.B. Dahlin, C. Meyer, L. Stenberg, T. Freier, C. Heimann, C. Barwig, L.F.V. Pinto, S. Raimondo, G. Gambarotta, S.R. Samy, N. Sousa, A.J. Salgado, A. Ratzka, S. Wrobel and C. Grothe, *Biomaterials*, 2013, **34**, 38, 9886.
174. A. Hadjizadeh and C.J. Doillon, *Journal of Tissue Engineering and Regenerative Medicine*, 2010, **4**, 7, 524.
175. G.R.D. Evans, K. Brandt, M.S. Widmer, L. Lu, R.K. Meszlenyi, P.K. Gupta, A.G. Mikos, J. Hodges, J. Williams, A. Gurlek, A. Nabawi, R. Lohman and C.W. Patrick, *Biomaterials*, 1999, **20**, 12, 1109.
176. J.J. Liu, C.Y. Wang, J.G. Wang, H.J. Ruan and C.Y. Fan, *Journal of Biomedical Materials Research, Part A*, 2011, **96A**, 1, 13.
177. L. Luo, L. Gan, Y. Liu, W. Tian, Z. Tong, X. Wang, C. Huselstein and Y. Chen, *Biochemical and Biophysical Research Communications*, 2015, **457**, 4, 507.
178. F. Xu, K. Zhang, P. Lv, R. Lu, L. Zheng and J. Zhao, *Materials Science and Engineering: C*, 2017, **70**, 2, 1132.
179. C.E. Schmidt, V.R. Shastri, J.P. Vacanti and R. Langer, *Proceedings of the National Academy of Sciences of the United States of America*, 1997, **94**, 17, 8948.
180. S. Catrina, B. Gander and S. Madduri, *European Journal of Pharmaceutics and Biopharmaceutics*, 2013, **85**, 1, 139.
181. K. Dalamagkas, M. Tsintou and A. Seifalian, *Materials Science and Engineering: C*, 2015, **65**, 425.
182. F.J. O'Brien, B.A. Harley, I.V. Yannas and L.J. Gibson, *Biomaterials*, 2005, **26**, 4, 433.
183. G.C.W. de Ruiter, M.J.A. Malessy, M.J. Yaszemski, A.J. Windebank and R.J. Spinner, *Neurosurgical Focus*, 2009, **26**, 2, E5.
184. K. Chwalek, Y. Dening, C. Hinuber, H. Brunig, M. Nitschke and C. Werner, *Materials Science and Engineering: C*, 2016, **61**, 466.
185. R.A. Koppes, S. Park, T. Hood, X. Jia, N. Abdolrahim Poorheravi, A.H. Achyuta, Y. Fink and P. Anikeeva, *Biomaterials*, 2016, **81**, 27.
186. S. Farah, D.G. Anderson and R. Langer, *Advanced Drug Delivery Reviews*, 2016, **107**, 367.
187. B. Tyler, D. Gullotti, A. Mangraviti, T. Utsuki and H. Brem, *Advanced Drug Delivery Reviews*, 2016, **107**, 163.
188. J. Cai, X. Peng, K.D. Nelson, R. Eberhart and G.M. Smith, *Journal of Biomedical Materials Research, Part A*, 2005, **75**, 2, 374.
189. M. Oudega, S.E. Gautier, P. Chapon, M. Fragoso, M.L. Bates, J.M. Parel and M.B. Bunge, *Biomaterials*, 2001, **22**, 1125.
190. C.O. Goulart, F.R. Pereira Lopes, Z.O. Monte, S.V. Dantas, A. Souto, J.T. Oliveira, F.M. Almeida, C. Tonda-Turo, C.C. Pereira, C.P. Borges and A.M.B. Martinez, *Methods*, 2015, **99**, 28.
191. M.C. Lu, Y.T. Huang, J.H. Lin, C.H. Yao, C.W. Lou, C.C. Tsai and Y.S. Chen, *Journal of Materials Science: Materials in Medicine*, 2009, **20**, 5, 1175.
192. S. Hui Hsu, S.H. Chan, C.M. Chiang, C. Chi-Chang Chen and C.F. Jiang, *Biomaterials*, 2011, **32**, 15, 3764.
193. D. Arslantunali, T. Dursun, D. Yucel, N. Hasirci and V. Hasirci, *Medical Devices: Evidence and Research*, 2014, **7**, 405.
194. W.F.A. Den Dunnen, B. Van der Lei, P.H. Robinson, A. Holwerda, A.J. Pennings and J.M. Schakenraad, *Journal of Biomedical Materials Research*, 1995, **29**, 6, 757.
195. G.C. de Ruiter, I.A. Onyeneho, E.T. Liang, M.J. Moore, A.M. Knight, M.J.A. Malessy, R.J. Spinner, L. Lu, B.L. Currier, M.J. Yaszemski and A.J. Windebank, *Journal of Biomedical Materials Research, Part A*, 2008, **84A**, 3, 643.

Gabriel Goetten de Lima, Shane Halligan, Luke Geever,
Maurice Dalton, Chris McConville and Michael J.D. Nugent

3 Controlled release of poorly soluble active ingredients from bioresorbable polymers

3.1 Introduction

Drug delivery is of significant importance to medical and healthcare sectors. It is a broad term that covers issues such as formulation, technology and the system for transporting active pharmaceutical ingredients (API) in the body [1]. The main objective of any drug-delivery device is to safely achieve the desired therapeutic result with minimum side effects. However, in some cases, such as chemotherapy for cancer, current treatment methods rely primarily on the use of conventional cytotoxic drugs that have an adverse side effect and only limited effectiveness. Many studies have indicated that these problems could be attributed to the lack of target specificity of state-of-the-art anti-tumour drugs. To overcome this hurdle the demand for drug delivery is growing and, in the US alone, the drug-delivery market was valued at >\$135 billion in 2015 as a result of the growth of 9% each year since 2010. The largest share is expected to be produced by oral, parenteral or injectable drugs [1].

Controlled drug delivery improves bioavailability (i.e., the proportion of a drug that enters the circulation when introduced into the body and so is able to have an active effect) by preventing premature degradation and enhancing uptake by controlling drug-release rates and reducing side effects by targeting the diseases or defective site. Since the first US Food and Drug Administration approved drug-delivery system, liposomal amphotericin in 1990, various drug-delivery systems have become available commercially to treat diseases ranging from cancer to fungal infections and muscular degeneration [2]. Polymers show improved pharmacokinetics, longer circulation times and more efficient tissue targeting in comparison with small-molecule drugs. They can be used as the polymeric drug, as a drug carrier, or be combined with small molecules such as proteins. The field is usually characterised by the term 'polymer therapeutics' and can be divided into five subclasses: polymeric drugs, polymer–drug conjugates, polymer–protein conjugates, polymeric micelles, and polyplexes.

If the polymeric material is not acting as a drug, it can be used as a drug carrier to reduce immunogenicity, toxicity or degradation while simultaneously improving circulation times and potentially having a passive targeting function. Polyethylene oxide (PEO) is a widely studied polymer in this field due to its biocompatibility, non-toxicity, water solubility and, most importantly, prolonged residence time in the body. PEO has been investigated to bypass normal physiological defences to protect chemical or biological conjugates by entrapping the PEO chains onto

https://doi.org/10.1515/9783110640571-003

conjugates to induce a stealth effect to decrease immunogenicity and increase the stability of the conjugates against enzyme hydrolysis [3].

The success of an efficient drug-delivery device hinges upon its ability to construct biocompatible carriers that allow high loading of drug molecules without premature release of the drug cargo before reaching the target organ [4]. Other factors to consider are that the carrier material should be compatible, and have high-loading capacity and a controlled release rate of drugs. For example, a novel polyethylene glycol 400 (PEG400)-mediated lipid nanoemulsion as a drug-delivery carrier for paclitaxel (PTX) was developed by Jing and co-workers [5]. They demonstrated that PEG400 could be used to carry an anticancer drug that is used for cancers of the breast, prostate gland and lung. Their investigation showed that a two-vial formulation comprising a PEG400 solution of the drug and a commercially available 20% lipid emulsion could be used to form paclitaxel-loaded lipid nanoemulsion (TPLE). Results demonstrated that TPLE could significantly reduce extraction of reticuloendothelial system organs and increase tumour uptake, which exhibited more potent antitumor efficacy in mice compared to conventional TPLE. TPLE did not cause haemolysis or an intravenous irritation response and showed the same cytotoxicity against HeLa cells as Taxol® [6].

Drug carriers are used to enhance therapeutic efficiency *via* encapsulation of active molecules. This encapsulation enhances the stability of drug molecules, improves targeting properties, and prolongs pharmacological activity *via* continuous local release of active molecules [7]. There is an array of drug-delivery devices, such as plasmids, viral, liposomes, micelles, dendrimers and nanoconjugates.

3.2 Poorly water-soluble drugs

Poorly water-soluble (hydrophobic) drugs have poor lipophilicity and, therefore, low stability, which are necessary constituents for desired bioavailability [8]. The study of drug-delivery systems for hydrophobic drugs is important because large amounts of active substances are difficult to formulate by normal approaches. This leads to undesirable distribution in the body due to their lack of solubility in the human body [9]. Enhancing the delivery technologies for hydrophobic drugs is very expensive because, according to van Hoogevest [8], various undesirable effects might occur: lack of stability in the solid state (chlorhexidine); extreme hygroscopicity (ibuprofen); extreme light sensitivity (nimodipine); and extremely low solubility (fenofibrate).

Although the lack of solubility in the human body does not prevent targeting the delivery of hydrophobic drugs [10], strategies have been developed to overcome the problems associated with solubility. Essentially, there are three main formulations for hydrophobic drug technologies: solubilised, amorphous and crystalline.

Solubilised formulations are generally used for oral drugs where the permeability through the intestinal epithelial membrane is not limited. The advantage of solubilised formulations is that dissolution of the drug is not necessary because the drug is already dissolved. However, precipitation of the drug may occur after release in the human body due to the dilution of the solubilising component if it is water-soluble; for an immiscible soluble component such precipitation does not occur unless the component remains intact [11, 12].

If the drug cannot solubilise for the method applied it is carried by an amorphous formulation of the drug, such as a solid dispersion. This process increases oral absorption but lacks stability, which leads to crystallisation [13, 14].

An amorphous formulation is the most robust; the drug needs to dissolve and is dependent on the crystalline phase and particle size. Nonetheless, to increase the dissolution rate, the particle size must be as small as possible. This leads to nano-sized particles, which are widely studied in this field [15].

Amorphous formulations can offer high free energy and lead to improved solubility [16]. However, the use of amorphous formulations is limited due to their thermodynamic instability, which causes higher chemical degradation and recrystallisation. In addition, solid dispersions often lead to poor reproducibility of physicochemical properties, difficulties in development of dosage forms and potential physical instability [17].

It is also necessary to analyse the route where the drug will be administered. Different administration routes might cause different problems due to the low dissolution of hydrophobic drugs.

The oral route is the most favoured because it can be being easily handled and is painless. Drugs administered *via* this route can also be produced in different dosages with low costs [18]. One of the reasons is that the gastrointestinal tract (GIT) has a high intestinal surface area and rich mucosal vasculature, which offers high absorption following bioavailability. GIT fluids dissolve the drug so it can be absorbed onto the intestinal mucosa and be delivered into the circulation. In this case, the aqueous solubility and intestinal permeability are key factors for this delivery route [19]. In the case of poorly water-soluble drugs, their behaviour results in low bioavailability and is highly sensitive to food intake [20]. Normally, to prevent some complications that may occur with the interaction of the drug and food intake, patients are suggested to follow food restrictions.

The parenteral route is the injection of drugs *via* subcutaneous, intra-arterial or intravenous routes. The drug is delivered into the blood circulation and allows rapid exertion of its pharmaceutical effect. This route is preferred for drugs for which the oral route might lead to low bioavailability. In addition, this route provides more predictable pharmacokinetic and pharmacodynamics profiles than that for oral administration [21]. However, despite the benefits of the pharmaceutical effects, drugs must pass some regulatory requirements and this is not feasible for poorly water-soluble drugs. To overcome this issue hydrophobic drugs are mixed with organic solvents, which might lead to a decrease in bioavailability and cause haemolysis and drug precipitation [22].

The pulmonary route is aimed towards diseases such as asthma. It leads to enhanced delivery of the pharmaceutical ingredient to tissue with a large surface area and thin epithelial barrier [19]. The particle size of the drug determines where it will be deposited into the lungs. For example, particles of size 1–3 μm are deposited deep within the lungs; those of 5 μm in the upper airways, and particles <1 μm are not deposited and instead exhaled after inspiration [19, 23]. Apart from this limitation of particle size, the pulmonary route is also limited by the small number of safe ingredients for this delivery route.

To overcome the issues associated with bioavailability and controlled release of poorly water-soluble drugs, the introduction of polymers for drug carriers has been investigated. In earlier studies, drug encapsulation was used merely for improvement of drug-delivery efficiency [24]. Currently, controlled-release mechanisms have been used to: target delivery of hydrophobic drugs *via* amorphous solid dispersions; improve API stabilisation through the inhibition of nucleation and crystallisation in the solid state; enhance effective solubility; minimise drug degradation; reduce drug toxicity; facilitate the control of drug uptake [25].

3.3 Mechanisms of drug release from biodegradable polymers

The mechanism of controlled release from biodegradable polymers can occur as diffusion through a rate-controlling membrane, osmosis, ion exchange or degradation. The advantage of drug delivery from biodegradable polymers is that these materials do not need to be removed from the body after drug delivery has been completed. Normally, bioresorbable polymers are usually considered because they degrade slowly and it is possible to control drug delivery for longer periods of time [26]. The mechanism of drug release from biodegradable polymers as well as their release kinetics is dependent on the choice of polymer and formulation method used [27, 28]. The hydrolytic cleavage of the polymer chains results in erosion of the polymer matrix, with two main mechanisms of degradation, bulk and surface, taking place [29, 30]. Whether bulk degradation or surface degradation is the dominant mechanism is dependent on the rate of water penetration and hydrolysis of the polymer backbone and is, therefore, influenced by the: rate of water diffusion through the polymer; size of the device; polymer porosity; reactivity of the functional groups of the polymer [31]. If the rate of water penetration is significantly higher compared with the rate of polymer hydrolysis, bulk degradation will be the dominant mechanism. This is because degradation of the polymer backbone will spread through the entire device as a result of the fast-penetrating water before surface degradation can take place. The opposite is true for surface degradation, with it being the dominant mechanism if degradation of the polymer matrix is faster than the rate of water diffusion into the device. This is due to surface erosion taking place before water has penetrated into the device [32].

Surface-degrading polymers can provide improved controlled-release kinetics compared with bulk-degrading polymers and thus have greater potential in drug-delivery applications. Polyanhydrides and polyorthoesters are two types of surface-degrading polymers that have been researched extensively for drug delivery. The chemical bonds (anhydride bonds and orthoesters, respectively) of their polymer backbone are hydrolysed faster than water can diffuse into the polymer, which results in surface degradation [33–35]. In theory, if the surface-degradation process is spread evenly and confined to the surface of the device or formulation, then drug release will be controlled by surface degradation and exhibit zero-order release kinetics [36]. However, actual drug release may deviate from zero-order release kinetics because it is dependent on the type, chemistry, molecular weight (MW), crystallinity, porosity and glass transition temperature of the polymer.

Bulk-degrading polymers such as polycaprolactone (PCL) and poly(lactic-co-glycolic acid) (PLGA) are the most commonly used biodegradable polymers in drug-delivery applications. The release rate, duration of release, and release profile (mono- or multi-phasic profiles) can be controlled by the degradation of the polymer. Therefore, a particular release profile can be achieved by selecting or modifying a polymer with an appropriate degradation rate or behaviour [37]. The rate of degradation and degradation behaviour of bulk-degrading polymers is influenced by several parameters:

1. Polymer composition – in PLGA, the glycolide group degrades faster than the lactide group, thus the release rate from PLGA can be adjusted by varying the ratio of these two groups [38, 39].
2. MW and polydispersity – the lower the MW of the polymer, the faster its degradation process will be. Therefore, the choice of a low-MW polymer will lead to increased drug release.
3. Polymer crystallinity – the more amorphous the polymer the faster rate of water penetration into the polymer and thus, the faster the polymer degrades [40].
4. The size of the drug-delivery device or formulation – larger devices can undergo autocatalytic degradation because the degradation products remain within the device and catalyse the degradation process further, thus increasing the degradation rate [41].

However, depending on the degradation rate, it could take longer for water to penetrate larger devices, which may slow down the degradation process. The nature and type or drug, drug loading, and porosity of the polymer can also influence release.

The process of degradation and thus drug release for bulk-degrading polymers involves four key steps:

1. The water or dissolution medium penetrates into the amorphous regions of the polymer and begins to disrupt the tertiary structure, which is stabilised by hydrogen bonds and van der Waal's forces.
2. The ester bonds in the polymer backbone are cleaved to generate carboxylic acids, which autocatalyse the hydrolysis process. The MW of the polymer will begin to decrease and the device or formulation will begin to lose its mechanical strength.
3. Cleavage of the ester bonds on the polymer backbone continues with a significant loss of the MW of the polymer as well as the mechanical strength of the device or formulation.
4. The oligomers are solubilised into the dissolution medium and the polymer breaks down into smaller fragments, which are subsequently hydrolysed into free acids.

The drug-release profile of bulk-degrading polymers is typically tri-phasic. An initial burst is followed by a period of zero-order drug release, controlled by a combination of drug diffusion through the polymer and degradation. The third phase is a rapid release of drug due to degradation of the polymer device or formulation [42–44].

3.4 Polymers for controlled release of poorly water-soluble drugs

3.4.1 Controlled release from water-soluble/bioresorbable hydrogels

Hydrogels are three-dimensional crosslinked polymeric networks that can swell and absorb large quantities of water or biological fluids to form a soft and elastic material that does not dissolve instantly. Swelling is due to the hydrophilic blocks of the group chains. With lower hydrophilicity, the polymer swells in water but, with a further increase in hydrophilicity, the polymer becomes water soluble [6, 45, 46]. Certain hydrogels can show responses to a stimulus in the environment such as volume, viscosity, pH, temperature or mechanical stress by incorporation of co-monomers into the network [45, 47, 48]. These hydrogels are termed 'stimuli-sensitive' or 'smart' hydrogels.

Hydrogels can be obtained naturally or by syntheses and have elicited considerable interest because, due to their properties, it is possible to obtain controlled drug delivery such as: coatings for the oral route and as dissolution and binding agents in capsules of drugs [49, 50]; encapsulation of cells for repairing and regenerating organs and tissues [46, 51]. Hydrogels have been investigated for targeted delivery of poorly water-soluble drugs. The incorporation of poorly water-soluble drugs improves their solubility and achieves longer periods of delivery, which increases the chances of intra-tumour uptake of drugs compared with free drugs [52].

Although the popular term 'hydrogel' refers to gels made by water-soluble stable polymers, they can be modified to become biodegradable. A common technique for the production of biodegradable hydrogels is to utilise low-MW polymers, which have a low degree of crosslinking. However, another approach is to produce hydrogels which are crosslinked with hydrolytically and enzymatically biodegradable chains. An example of this is the use of hydroxyethyl methacrylate which is crosslinked with PCL-based crosslinking agent rendering the entire hydrogel biodegradable [53].

Early studies on the drug release of biodegradable hydrogels [26] showed interesting behaviour. The drug will be in contact with water, so the drug properties are important and its solubility must be addressed. This strategy is usually employed for macromolecular drugs, which are entrapped in the gel network until the gel is degraded.

The pathways for biodegradability or dissolution may occur *via* enzymatic means, hydrolytic means, or *via* the environment. For drug delivery, degradation is not always desirable due to the timescale and location for the drug-delivery device.

An amorphous solid dispersion is the most used formulation when incorporating poorly water-soluble drugs onto hydrogels. Crosslinked hydrogels, compared with water-soluble polymers for poorly water-soluble drugs, have considerable advantages. Sun and Lee [54] stated that, when using amorphous solid dispersions on water-soluble polymers, a highly supersaturated environment is created. This results in recrystallisation of the dissolved drug, which reduces the drug concentration and supersaturation limit of the solid dispersion. Conversely, crosslinked hydrogels ensure that the drug is 'tangled' and protected by the hydrogel layer. This helps control drug delivery by reducing diffusion to the external core and preventing supersaturation in the environment. This could lead to a critical supersaturation above which spontaneous nucleation and crystallisation would occur (Figure 3.1).

However, hydrogels have limited capacity for loading of poorly water-soluble drugs. Two methods have been reported [52, 55]: incorporation of polymeric micelles onto hydrogels that can loco-regionally deliver multiple drugs in a single dose [56]; and electrospinning to produce nanofibres with poorly water-soluble drugs to enhance the solubility, drug release and effectiveness of the active ingredient [55].

3.4.2 Controlled release from polymeric micelles

As discussed before, nanoparticles help to improve poor solubility. In this regard, polymer micelles have been used in multifunctional-based delivery systems for poorly water-soluble drugs. Polymeric micelles consist of a hydrophobic and hydrophilic block domain that are self-assembled and which can form a 'core-shell' structure due to the interaction of the hydrophobic and hydrophilic parts named as 'micelles' [57–59]. The interior part, the hydrophobic block, protects the drug and the hydrophilic part confers solubility [59]. Micelle formation follows after the

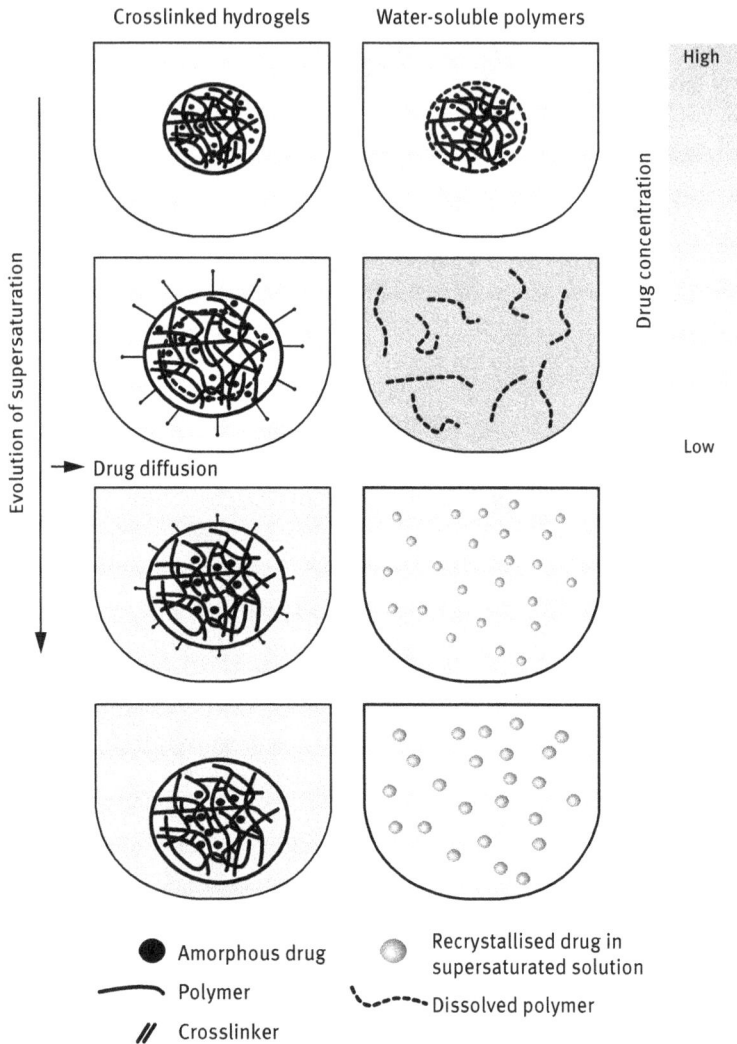

Figure 3.1: Drug-release mechanism of an amorphous drug in crosslinked hydrogels and water-soluble polymers. Reproduced with permission from D.D. Sun and P.I. Lee, *Acta Pharmaceutica Sinica: B*, 2014, **4**, 26. ©2013, Elsevier [54].

colloid aggregates, and the concentration range at which these chemical structures are formed is above the critical micelle concentration [60].

The most used polymer to form this structure is polyethylene glycol (PEG) at a MW range of 2–15 kDa. It can stabilise the micelle structure and interactions with cells and proteins. Subsequently, a hydrophobic block consisting usually of PCL [61], poly(b-benzyl-l-aspartate) [62] or polypropylene oxide (PPO) [63] acts as a shell to protect the hydrophobic drug from making contact with the aqueous solution and hindering enzymatic degradation of bioactive molecules [64]. Use of these polymers is based primarily due to their non-toxicity, biodegradability and biocompatibility. Also, PEG can decrease opsonin adhesion in plasma and increase the circulation time of the structures in the human body [65]. The advantage of using polymeric micelles compared with micelles is that they have a very low critical micelle concentration, which results in relatively stable systems which are not easily dissociable *in vivo* [66, 67]. This stability ensures the controlled administration of the pharmaceutical ingredient for longer periods of time.

In general, it is possible to prepare polymeric micelles with many copolymers but the choice becomes limited because of biocompatibility and biodegradability [68]. The most used block copolymers are the ones characterised by their hydrophobic blocks, such as polyethers, polyamides and polyesters.

Drug loading onto polymeric micelles can be influenced by the hydrophobic effect and the interaction between the micelle blocks. Several studies [65, 69] have shown that the interaction between hydrophobic and hydrophilic parts is significant. For example, PEG-*b*-PCL polymeric micelles that have a longer hydrophobic block improve drug capabilities [70], but the partition coefficient can deteriorate with a longer PEG chain. Miscibility is the main factor that can improve drug loading and the initial amount of drug influences the entrapment efficiency which is also related to the aggregation number of the polymer block [71, 72].

The advantages of using polymeric micelles for poorly water-soluble drugs are: their very small size (which helps to target drug-administration routes such as blood vessels); high-structural stability due to their core-shell structure [73], which helps to improve bioavailability; large amount of drug loading and water solubility due to their inherent layer of hydrophilic/hydrophobic parts [74]; incorporation of various diagnostic agents.

3.5 Recent strategies for poorly water-soluble drugs

3.5.1 Hydrogels/polymeric micelles

Recent research shows that incorporation of micelles as a hydrophobic block into the hydrogel structure can resemble the behaviour of semicrystalline polymers which contains a mixture of crystalline and amorphous regions [75]. This balance can provide water absorption, fluid flow, and controlled release based on the hydrophilic block, whereas the hydrophobic segment can provide the strength, tear and shear resistance [76]. This structure is quite advantageous for poorly water-soluble drugs. Ju and co-workers [77] demonstrated that when using PTX (an anticancer agent with poor water solubility), micellar encapsulation onto the hydrogel reduced the toxicity of PTX and the whole system demonstrated prolonged drug release (>15 days).

A block of polymeric micelles in the hydrogel structure helps solubilise poorly water-soluble drugs and acts as a 'shell' for transport and controlled delivery [78]. A system for oral delivery of the anticancer agent docetaxel was developed to target the drug onto the intestinal tract while improving its solubility. Polymeric micelles were mixed with docetaxel, which showed better solubility and permeability. In addition, encapsulation of the micelle block helped prevent identification by proteins that could cause efflux and change delivery within the body.

Three main methods can be used to incorporate hydrogels containing micelles, as shown below:

1. Hydrogels with distributed micelles – using this method, a mixture of micelles and a hydrogel matrix is formed [69]. Hydrogels with micelles have a complex structure that can host hydrophilic and hydrophobic drugs. Anirudhan and co-workers showed that the release of a hydrophilic drug (tramadol hydrochloride) was controlled mainly by the hydrogel and release of a hydrophobic drug (cefixime trihydrate) was controlled mainly by the polymeric micelle [79].

2. Hydrogels with integrated micelles – using this method, it is possible to obtain two type of structures:
   - Dispersed micelles anchored in crosslinked chains – very insightful work by Yom-Tov and co-workers [79] used pluronic micelles in a hydrogel while anchoring most of their molecules to the surrounding network through their end groups. This led to anchored molecules that facilitated the crosslinking between the hydrogel and stabilised the network.
   - Micelles formed by hydrophobic segments located directly on the gel – Inoue and co-workers [80] grafted oligo(methyl methacrylate) as a hydrophobic block onto hydrophilic polyacrylic acid (PAA) and used model hydrophobic drugs. The hydrophobic drug was released slowly from the hydrogels compared with an ungrafted-PAA hydrogel. The authors hypothesised that this was due to favoured absorption of the hydrophobic drug onto the hydrophobic block and lack of interaction, which led to faster drug release for hydrophilic drugs.

3. Hydrogels with micellar crosslinks – in this method, the polymeric micelles form a chemical or physical crosslinking network with the hydrogel. Missirlis and co-workers [81] used inverse emulsion photocopolymerisation with PEG diacrylate to produce pluronic-based nanoparticles. The hydrophobic blocks within these biodegradable nanoparticles solubilised a poorly water-soluble drug, doxorubicin, up to almost 10% (w/w).

Loading of a hydrophobic drug onto hydrogel-polymeric micelles involves mixing a solution of micelles with the drug followed by the incorporation onto the hydrogel. These processes include: simple equilibrium (dissolution of the drug into a solution of pre-formed micelles); chemical conjugation; dialysis; oil–in–water emulsions;

solution casting; and freeze-drying. For the detailed process, the reader should find [60, 64] relevant.

Figure 3.2 shows how micelles are embedded in a hydrogel structure. In this example, the polymeric micelles entrap the hydrophobic drugs acting as a core-shell for the drug, following immersion mixing with the hydrogel. The representation shows well-aligned micelles between the hydrogel. However, as explained above, one can obtain three major structures with very good bioavailability and low toxicity without altering the biodegradability of the hydrogel for targeted delivery of the drug.

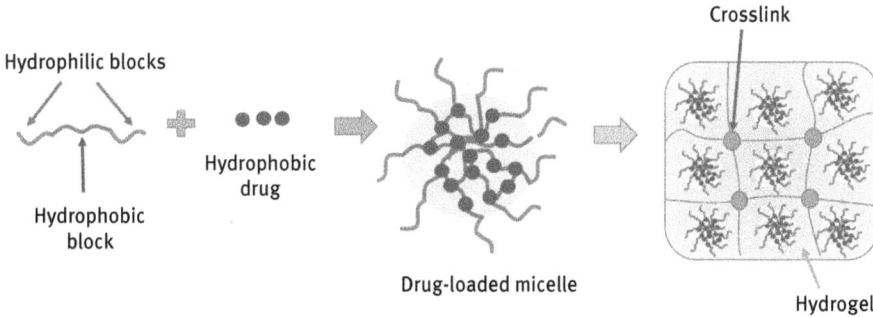

Figure 3.2: Nanoscale drug-entrapment strategy in hydrogels (schematic).

3.5.2 Nanocomposite hydrogels

Due to the deficiency of loading for hydrophobic drugs, hydrogels have been produced at the nanoscale [82]. Gwak and co-workers [83] prepared a hydrogel-inorganic hybrid material with agarose mixed with zinc basic salt, and the results showed sustained release. Apart from micelles, biodegradable cyclodextrin (CD) hydrogels have also been studied for complex poorly water-soluble drugs [84] to increase their solubility. The incorporation of CD into hydrogels does not only maintain the swelling properties of the hydrogel but the hydrophobic interior can facilitate the capture and sustained release of a hydrophobic drug. The process of CD incorporation is similar to that for micelles [82], but the drug can be added onto the hydrogel during or after gel crosslinking.

Gao and co-workers [85] produced a thermosensitive hydrogel-based poly (ε-caprolactone)–polyethylene glycol–poly(ε-caprolactone) (PCEC), which has the potential for preventing formation of postoperative adhesions. They used ring-opening polymerisation of the PCEC copolymer to dissolve it into PCEC micelles, which resulted in spheres of mean size 25 nm (Figure 3.3A). This polymer was thermosensitive and at body temperature the PCEC micelles were converted into a hydrogel (Figure 3.3B). The results showed no signs of

(A)

Self-assembly

Hydrophobic drug

PCL-PEG-PCL

PCL-PEG-PCL

(B)

Small micelles at low temperature

Large micelles at high temperature

Hydrogel forming at gel temperature

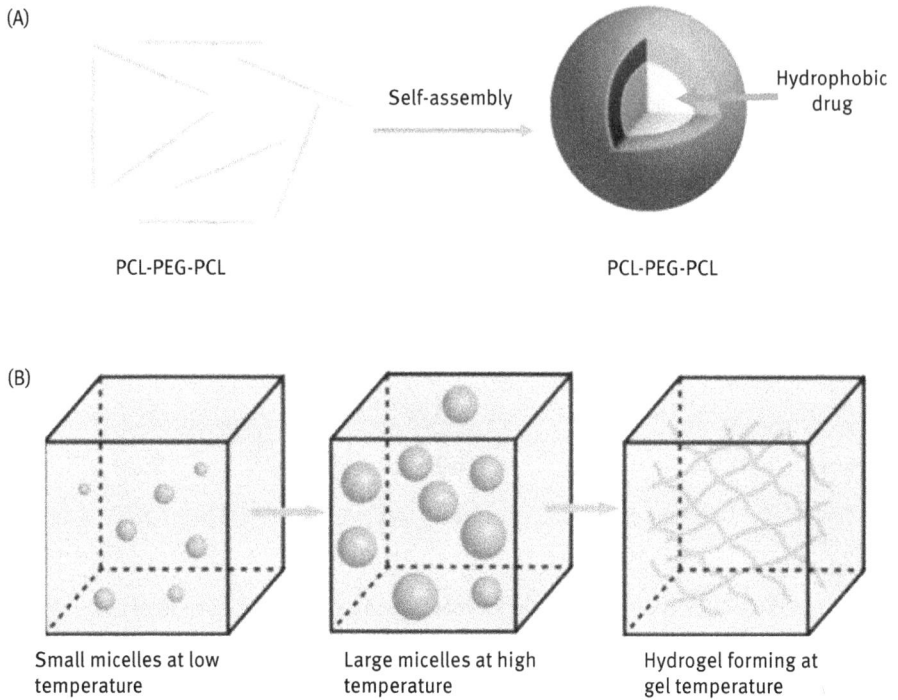

Figure 3.3: A thermosensitive PCEC hydrogel (schematic). (A) Assembling of PCEC triblock copolymers and (B) due, to the sensitivity of these micelles, they became larger with increasing temperature, and a hydrogel was formed when the temperature reached the gelation temperature. Reproduced with permission from X. Gao, X. Deng, X. Wei, H. Shi, F. Wang, T. Ye, B. Shao, W. Nie, Y. Li, M. Luo, C. Gong and N. Huang, *International Journal of Nanomedicine*, 2013, **8**, 2453. ©2013, Dove Medical Ltd [85].

adhesion *in vivo* and of total degradation after 6 weeks. According to Gao and co-workers, this system can also be used for delivery of two drugs.

3.6 Stimuli-responsive hydrogels

During the 1970s, hydrogel research shifted from water-swollen networks to hydrogels that were capable of responding to changes in environmental conditions [86]. In the past two decades, these stimuli-responsive hydrogels have received increasing attention in relation to several biomedical applications, such as drug-delivery systems and tissue engineering [87, 88]. Thus, stimuli-responsive hydrogels have been suggested to be superior to conventional hydrogels as sensors [89]. Stimuli-responsive hydrogels can react to minor changes in physical or chemical environments. One of the most promising applications for a stimuli-responsive hydrogel is drug delivery. This type of delivery allows the drug to increase its concentration in a specific organ or tissue, which can improve the efficiency and safety of some medicines that are already on the market. The development and success of targeted drug-delivery systems has led to novel strategies, such as delivery of anticancer drugs and gene therapy [90]. This helps the efficacy of the drug by reducing side effects. Treatment of most diseases is reliant on effective, safe, targeted drug-delivery systems [90]. Several types of stimuli-responsive hydrogels can react to a number of different environmental conditions. These can be classified into two main categories, which are displayed in Table 3.1 [88, 91, 92].

Table 3.1: Different types of stimuli.

Physical stimuli	Chemical stimuli
Temperature	pH
Mechanical stress	Ionic
Magnetic	–
Electric	–

The most frequently used stimuli are pH and temperature. Among all stimuli-responsive hydrogels, temperature-responsive hydrogel systems have received most interest because temperature is a vital physiological factor in the body. Some diseases manifest themselves by a change in temperature [93–96].

Dual-sensitive hydrogels can also be produced; this is where two stimuli-responsive monomers are bonded together to create a hydrogel that can respond to both stimuli. For example, combining pH and temperature-responsive monomers together will result in pH–temperature-sensitive hydrogels, where the phase-separation temperature will depend on the pH of the medium. Combining a pH monomer to a temperature-responsive hydrogel benefits an oral drug-delivery system. That is, the pH hydrogel will protect the drug in the presence of an acid pH (e.g., stomach) and deliver it to an area of alkaline pH (e.g., small intestine).

3.6.1 Hydrogels that are responsive to temperature

Temperature-responsive hydrogels are the most commonly studied due to their physiological importance. Temperature-responsive hydrogels contain hydrophilic and hydrophobic segments. If the temperature is within the range of phase transition, the balance between the hydrophilic and hydrophobic segments is altered and phase separation or sol-gel transition can follow [97]. Temperature-responsive hydrogels that can undergo a sol-gel transition have been shown to have applications in biomedical fields, particularly in drug delivery and injectable-tissue engineering [97–99]. These are solutions of hydrogels that can be injected into any tissue, organ or body cavity in a minimally invasive manner prior to sol-gel transition [86, 100].

Hydrogels that display a sol-gel transition around room temperature to body temperature are predicted to be the most beneficial [101]. Some hydrogels can comprise drugs or cells and such systems aid in delivery to a desired site. During a sol-gel transition, all of the pharmaceutical agents in the system can be released in a controlled manner [102]. Hydrogels that can undergo sol-gel transition offer several advantages over systems that must be formed into their final shape before implantation [86].

A temperature-sensitive hydrogel shows a thermally-induced phase transition in solutions. The phase behaviour of temperature-sensitive hydrogels is governed by the Gibbs' free energy of mixing (Equation 3.1):

$$\Delta G_m = \Delta H_m - T\Delta S \tag{3.1}$$

where ΔH_m and $T\Delta S$ are the enthalpy and the entropy of mixing, respectively.

Two components will mix only if the Gibbs' free energy of mixing is negative: ΔG_m < 0. Upon heating, the entropy of mixing can dominate and lead to decreasing miscibility with increasing temperature. A typical phase diagram consists of two-phase boundaries. The lower boundary corresponds to thermally-induced mixing (known as the upper critical solution temperature) and the upper boundary corresponds to thermally-induced demixing [known as the lower critical solution temperature (LCST)]. Temperature-responsive hydrogels, which become insoluble upon heating, have a LCST in which the hydrophilic and hydrophobic balance changes. Hydrogel systems that exhibit LCST behaviour have gained significant attention in biomedical applications in recent years. The change in temperature affects the hydrogen bonding of the polymer molecules, which results in a volume-phase transition [103].

One of the most renowned temperature-responsive hydrogels is poly(N-isopropylacrylamide) (PNIPAAm), which exhibits a phase transition of ≈32°C in aqueous solution. The aim of researching temperature as a release trigger is due to controlled drug release by precisely controlling the temperature sensitivity of the hydrogel to the environmental temperature of the target site. Temperature-responsive hydrogels have become attractive candidates for designing nano-vehicles to target specific tissues. The temperature range at which these nano-vehicles should release is 37–42 °C [104].

Low-MW poly(N-vinylcaprolactam) (PNVCL) is known for its superior biocompatibility, solubility, thermosensitivity and having non-ionic and non-toxic features. Moreover, PNVCL has a similar LCST range compared with PNIPAAm, which is between the physiological range of 32 and 38 °C [105]. The LCST behaviour of PNVCL is sensitive to alterations in the polymer concentration, the MW of the polymer and also the composition of the solution [106]. PNVCL sensitivity to these alterations allows for 'tuned' LCST behaviour.

3.6.2 Hydrogels that are responsive to pH

Hydrogels that are responsive to pH are also very popular stimuli. The pH sensitivity of a hydrogel results from weak acid or weak base functionality on the polymer backbone, which accepts or releases protons depending upon external conditions [107]. Thus, at certain pH values, they can dissociate in aqueous solutions to form polyelectrolytes. Commonly used weak acids are acrylic acid and methacrylic acid [49, 108]. A frequently used weak base is dimethylaminoethyl methacrylate. The pH at which hydrogels show volume changes depends on the type of weak acid or base used. Some hydrogels can contain weak acid functionality; they will swell as the pH of the medium increases. Hydrogels that are pH-sensitive are commonly used to develop controlled-release formulations for oral administration [109]. Certain body tissues have different pH values, such as the stomach, which is quite different from the neutral pH in the intestine, and more changes occur within various body tissues. Chronic wounds can have a pH range between 7.4 and 5.4, and cancer tissue has been reported to be acidic extracellularly [103]. Cancer tissue usually has a pH <7 and the pH of heathy body tissue is 7.2–7.4. Tumours with low pH result primarily from a high glycolysis rate, which can produce lactic acid. This low pH benefits tumour cells and promotes invasive cell growth. The low pH in the tumour provides a tissue-specific stimulus that may be exploited to target applications. pH-responsive hydrogels can be tailored to carry, deliver, and control the release of a therapeutic agent in cancer tissue (Figure 3.4 [110]).

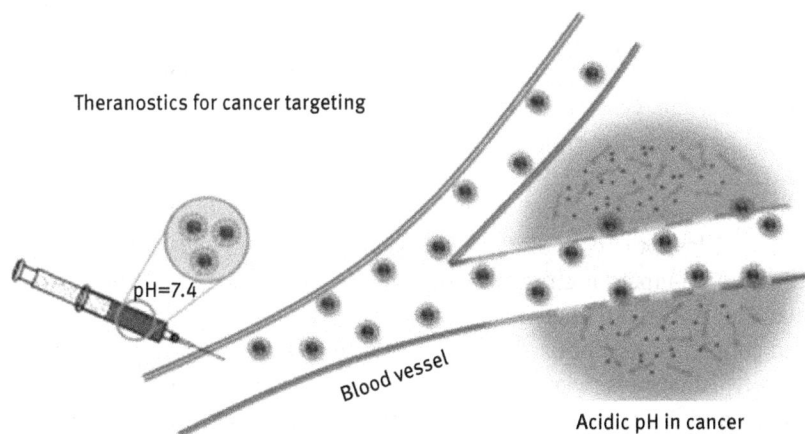

Figure 3.4: pH responsive hydrogels can be tailored to carry, deliver and control the release of a therapeutic agent for cancer treatment due to the reduction in pH of the tissue at the tumour site. Reproduced with permission from G.H. Gao, Y. Li and D.S. Lee, *Journal of Controlled Release*, 2013, **169**, 180. ©2013, Elsevier [110].

3.6.3 Combining pH and temperature-responsive hydrogels

Hydrogels can be sensitive to both pH and temperature. The presence of pH-sensitive co-monomers, which are charged, increases the LCST of the thermosensitive co-monomers due to an increase in the hydrophilicity of the polymer. pH and temperature-sensitive hydrogels can undergo volume changes in response to changes in temperature and pH. The magnitude and location of temperature-induced gel collapse can be altered significantly by changing the pH of the solution. This allows tailoring of a thermosensitive hydrogel to meet specific process or system needs. These unique characteristics are of great interest in drug delivery [111] and cell encapsulation [112].

An injectable pH and temperature-sensitive biodegradable hydrogel poly (β-amino ester urethane) and triblock poly(ε-caprolactone lactide) (PCLA)–PEG–PCLA was used to enhance sustained release by dual ionic interactions. The injectable hydrogel suppressed the initial burst release and extended the release period for 13 days *in vitro* and 5 days *in vivo* [113].

3.7 Applications of stimuli-responsive hydrogels

Applications that require the delivery system not to be degradable, such as in oral drug delivery, do not offers problems. However, this becomes a serious limitation in other applications. One major disadvantage of temperature-responsive hydrogels such as PNIPAAm and PNVCL is that they are not degradable. Thus, recent research has focused on development of the biodegradable ability of these hydrogels. For a hydrogel to be biodegradable, the presence of hydrolytic or proteolytic labile bonds in their backbone is essential to allow it to degrade. Several research teams have attempted to incorporate a biodegradable backbone to enhance the biodegradable properties of these hydrogels [114].

Rejinold and co-workers [115] developed curcumin-loaded chitosan (CS)-*g*-PNVCL nanoparticles for cancer therapy. This hydrogel system had biodegradable, biocompatible and temperature-responsive properties. It was extremely efficacious against cancer cells with minimal toxicity to normal cells *in vitro*.

Gan and co-workers synthesised novel temperature-sensitive biodegradable hydrogels from N-isopropylacrylamide and two biodegradable crosslinkers: PCL dimethacrylate and *bis*(acryloyl) cystamine. At physiological pH, the hydrogels could be biodegraded slowly in glutathione at 37 °C [116].

Kokuryo and co-workers [117] measured the release of anticancer drugs by dual-sensitive hydrogels (pH-temperature). They noted a decrease in tumour size after treatment.

3.7.1 Injectable hydrogels in drug delivery

Hydrogels that can be formed *in situ* can be clear polymer solutions prior to adminis-tration, and undergo a sol-gel transition in response to changes in stimuli. Physically crosslinked injectable hydrogels are reversible networks and can be developed by varying the environmental stimuli. Several appealing biomedical applications have been proposed for hydrogels that can be formed *in situ*, particularly in the areas of drug delivery and cell culture. Due to their unique advantages, such as minimal inva-sion, lack of organic solvents or photoinitiators, site specificity, and ability to deliver hydrophobic/hydrophilic drugs, stimuli-sensitive hydrogels are injectable. For exam-ple, drugs or cells can be encapsulated with *in situ*-forming hydrogels solutions at low temperature, and the mixed solutions rapidly undergo sol-gel transition after injection into the body [118]. This large gain in entropy is caused by the release of bound water molecules from the hydrophobic segments at sol-gel transition. The polymer accumulates and the phase-separated depot then slowly releases the drug by dissolution and diffusion of the drug.

In drug delivery, the release of a drug could be in response to a temperature in-crease that makes the temperature-responsive hydrogel undergo phase transition. The temperature at the target site is slightly higher (e.g., in solid tumours) than normal body temperature. By adjusting the sol-gel transition of the temperature-responsive hydrogel between the body temperature and the higher temperature of the tumour, it is possible for the drug-delivery system to accumulate within the tumour [119].

Prabaharan and co-workers entrapped self-assembled stable micelles of PNVCL-*b*-PEG block copolymers coupled with folic acid as an anticancer drug car-rier. The diameter of these micelles in a dehydrated state was ≈150 nm. At 37 °C, the release profile of these polymeric micelles showed a slower and more controlled re-lease of entrapped 5-fluorouracil (5-FU) than that at 25 °C. 5-FU-loaded micelles did not induce remarkable cytotoxicity. However, they showed a cytotoxic effect against 4T1 mouse mammary carcinoma cells due to the availability of loaded anti-cancer drugs delivered to the inside of cancer cells by a folate receptor-mediated endocytosis process [120, 121]

Yerriswamy and co-workers prepared PNVCL-*co*-VAc microspheres by free-radical emulsion polymerisation in the presence of 5-FU [122]. The authors reported higher release rates of 5-FU for the formulations prepared with higher amounts of N-vinylcaprolactam (NVCL) than formulations prepared using lower amounts of NVCL. Slower drug release was observed from formulations prepared with a lower amount of NVCL, which was attributed to the hydrophilic nature of the drug and NVCL in the copolymer [122].

Prabaharan and co-workers used biodegradable CS-*g*-PNVCL for the controlled re-lease of the hydrophobic drug ketoprofen. These were stabilised by crosslinking with a solution of sodium tripolyphosphate. The release behaviour was influenced by the pH and temperature of the medium. At pH 7.4 and 37 °C, CS-g-PNVCL showed a compact structure with reduced pore size and strong hydrophobic interactions with drug molecules, resulting in a slow and steady release of the drug from the system [123].

Thermogelling hydrogels can be incorporated into contact lenses. Thermogelling hydrogels are becoming popular in eye treatment because conventional eye drops move away quickly from the surface of the eye, resulting in poor bioavailability. To improve such poor bioavailability, Cho and co-workers developed a temperature-sensitive biodegradable hexanoyl glycol chitosan (HGC) as a carrier for topical drug delivery to the eye. The researchers found that controlling the amount of hexanoyl added to the glycol CS could control the thermogelling behaviour for glaucoma therapy. *In vivo* experiments demonstrated HGC to be maintained on the periocular surface for a longer period of time due to its increased viscosity at body temperature [124].

3.8 Summary

This chapter discussed the various difficulties associated with poorly water-soluble drugs for targeted delivery and bioavailability. Various strategies have been studied to improve it, but this field needs further work. The ability of drug holding and shell encapsulation offer hope in terms of attachment of cells and biocompatibility. In addition, targeted delivery with poorly water-soluble drugs will be able, in the near future, to localise where the drug needs to be delivered independent of the route chosen. Currently, pH and temperature-hydrogels allow a new future in biomedical applications using altered sol-gel transition. Although there have been achievements in pH and temperature-responsive injectable hydrogels, several challenges for clinical applications remain.

References

1. E. Skorupska, A. Jeziorna, S. Kazmierski and M.J. Potrzebowski, *Solid State Nuclear Magnetic Resonance*, 2014, **57**, 2.
2. Y. Zhang, H.F. Chan and K.W. Leong, *Advanced Drug Delivery Reviews*, 2013, **65**, 104.
3. S. Stolnik, L. Illum and S.S. Davis, *Advanced Drug Delivery Reviews*, 1995, **16**, 195.
4. I.I. Slowing, J.L. Vivero-Escoto, C-W. Wu and V.S-Y. Lin, *Advanced Drug Delivery Reviews*, 2008, **60**, 1278.
5. X. Jing, L. Deng, B. Gao, L. Xiao, Y. Zhang, X. Ke, J. Lian, Q. Zhao, L. Ma, J. Yao and J. Chen, *Nanomedicine: Nanotechnology, Biology, and Medicine*, 2014, **10**, 371.
6. Y. Jiang, J. Chen, C. Deng, E.J. Suuronen and Z. Zhong, *Biomaterials*, 2014, **35**, 4969.
7. K. Miladi, S. Sfar, H. Fessi and A. Elaissari, *International Journal of Pharmaceutics*, 2013, **445**, 181.
8. P. van Hoogevest, X. Liu and A. Fahr, *Expert Opinion on Drug Delivery*, 2011, **8**, 1481.
9. A. Fahr and X. Liu, *Expert Opinion on Drug Delivery*, 2007, **4**, 403.

10. A. Fahr, P. van Hoogevest, S. May, N. Bergstrand and M.L.S. Leigh, *European Journal of Pharmaceutical Sciences*, 2005, **26**, 251.
11. R.G. Strickley, *Pharmaceutical Research*, 2004, **21**, 201.
12. C.W. Pouton, *European Journal of Pharmaceutical Sciences*, 2006, **29**, 278.
13. Y. Kawabata, K. Wada, M. Nakatani, S. Yamada and S. Onoue, *International Journal of Pharmaceutics*, 2011, **420**, 1.
14. C. Brough, and R.O. Williams, *International Journal of Pharmaceutics*, 2013, **453**, 157.
15. C.M. Keck and R.H. Müller, *European Journal of Pharmaceutics and Biopharmaceutics*, 2006, **62**, 3.
16. P. Gupta, G. Chawla and A.K. Bansal, *Molecular Pharmaceutics*, 2004, **1**, 406.
17. P.H.L. Tran, T.T.D. Tran, J.B. Park and B.J. Lee, *Pharmaceutical Research*, 2011, **28**, 2353.
18. F. Gabor, C. Fillafer, L. Neutsch, G. Ratzinger and M. Wirth in *Drug Delivery*, Springer, Amsterdam, The Netherlands, 2010, p.345.
19. S. Bosselmann and R.O. Williams, III., in *Formulating Poorly Water Soluble Drugs*, Springer, New York, NY, USA, 2012, p.1.
20. G.L. Amidon, H. Lennernäs, V.P. Shah and J.R. Crison, *Pharmaceutical Research*, 1995, **12**, 413.
21. C.B. Packhaeuser, J. Schnieders, C.G. Oster and T. Kissel, *European Journal of Pharmaceutics and Biopharmaceutics*, 2004, **58**, 445.
22. S.H. Yalkowsky, J.F. Krzyzaniak and G.H. Ward, *Journal of Pharmaceutical Sciences*, 1998, **87**, 787.
23. J. Heyder, J. Gebhart, G. Rudolf, C.F. Schiller and W. Stahlhofen, *Journal of Aerosol Science*, 1986, **17**, 811.
24. A.P. Simonelli, S.C. Mehta and W.I. Higuchi, *Journal of Pharmaceutical Sciences*, 2016, **58**, 538.
25. J. Liu, and L. Li, *European Journal of Pharmaceutical Sciences*, 2005, **25**, 237.
26. H. Park, K. Park and W.S.W. Shalaby in *Biodegradable Hydrogels for Drug Delivery*, CRC Press, Boca Raton, FL, USA, 2011, p.1.
27. C. McConville, P. Tawari and W. Wang, *International Journal of Pharmaceutics*, 2015, **494**, 73.
28. I. Zembko, I. Ahmed, A. Farooq, J. Dail, P. Tawari, W. Wang and C. Mcconville, *Journal of Pharmaceutical Sciences*, 2015, **104**, 1076.
29. D.Y. Arifin, L.Y. Lee and C-H. Wang, *Advanced Drug Delivery Reviews*, 2006, **58**, 1274.
30. K.L. Poetz, H.S. Mohammed, B.L. Snyder, G. Liddil, D.S.K. Samways and D.A. Shipp, *Biomacromolecules*, 2014, **15**, 2573.
31. F. von Burkersroda, L. Schedl and A. Göpferich, *Biomaterials*, 2002, **23**, 4221.
32. S. You, Z. Yang and C-H. Wang, *Journal of Pharmaceutical Sciences*, 2016, **105**, 1934.
33. A. Göpferich, *Biomaterials*, 1996, **17**, 103.
34. J. Tamada and R. Langer, *Journal of Biomaterials Science, Polymer Edition*, 1992, **3**, 315.
35. B. Wei, Y. Tao, X. Wang, R. Tang, J. Wang, R. Wang and L. Qiu, *ACS Applied Materials & Interfaces*, 2015, **7**, 10436.
36. J. Heller and J. Barr, *Biomacromolecules*, 2004, **5**, 1625.
37. L.L. Lao, N.A. Peppas, F.Y.C. Boey and S.S. Venkatraman, *International Journal of Pharmaceutics*, 2011, **418**, 28.
38. S. Li, *Journal of Biomedical Materials Research*, 1999, **48**, 342.
39. X.S. Wu and N. Wang, *Journal of Biomaterials Science, Polymer Edition*, 2001, **12**, 21.
40. M. Vert, S. Li and H. Garreau, *Journal of Controlled Release*, 1991, **16**, 15.
41. S. Li, H. Garreau and M. Vert, *Journal of Materials Science: Materials in Medicine*, 1990, **1**, 198.
42. B.S. Zolnik, P.E. Leary and D.J. Burgess, *Journal of Controlled Release*, 2006, **112**, 293.
43. N.S. Berchane, K.H. Carson, A.C. Rice-Ficht and M.J. Andrews, *International Journal of Pharmaceutics*, 2007, **337**, 118.
44. Q. Xu and J.T. Czernuszka, *Journal of Controlled Release*, 2008, **127**, 146.

45. G.G. de Lima, D. Kanwar, D. Macken, L. Geever, D.M. Devine and M.J.D. Nugent in *Handbook of Polymers for Pharmaceutical Technologies*, Eds., V.K. Thakur and M.K. Thakur, John Wiley & Sons, Inc., New York, NY, USA, 2015, p.1.
46. A.S. Hoffman, *Advanced Drug Delivery Reviews*, 2012, **64**, 18.
47. S. Chaterji, I.K. Kwon and K. Park, *Progress in Polymer Science*, 2007, **32**, 1083.
48. Y. Samchenko, Z. Ulberg and O. Korotych, *Advances in Colloid and Interface Science*, 2011, **168**, 247.
49. M. Canillas, G.G. de Lima, M.A. Rodríguez, M.J.D. Nugent and D.M. Devine, *Journal of Polymer Science, Part B: Polymer Physics*, 2015, **54**, 761.
50. T.G. Park, *Biomaterials*, 1999, **20**, 517.
51. N.A. Peppas, P. Bures, W. Leobandung and H. Ichikawa, *European Journal of Pharmaceutics and Biopharmaceutics*, 2000, **50**, 27.
52. M. McKenzie, D. Betts, A. Suh, K. Bui, L.D. Kim and H. Cho, *Molecules*, 2015, **20**, 20397.
53. S. Atzet, S. Curtin, P. Trinh, S. Bryant and B. Ratner, *Biomacromolecules*, 2008, **9**, 3370.
54. D.D. Sun, and P.I. Lee, *Acta Pharmaceutica Sinica: B*, 2014, **4**, 26.
55. U. Paaver, J. Heinämäki, I. Laidmäe, A. Lust, J. Kozlova, E. Sillaste, K. Kirsimäe, P. Veski and K.Kogermann, *International Journal of Pharmaceutics*, 2015, **479**, 252.
56. S.A. Abouelmagd, B. Sun, A.C. Chang, Y.J. Ku and Y. Yeo, *Molecular Pharmaceutics*, 2015, **12**, 997.
57. V.P. Torchilin, *Journal of Controlled Release*, 2001, **73**, 137.
58. P. Mukerjee, *Journal of Pharmaceutical Sciences*, 1971, **60**,1531.
59. A.S. Mikhail and C. Allen, *Journal of Controlled Release*, 2009, **138**, 214.
60. B.P.K. Reddy, H.K.S. Yadav, D.K. Nagesha, A. Raizaday and A. Karim, *Journal of Nanoscience and Nanotechnology*, 2015, **15**, 4009.
61. H. Ai, C. Flask, B. Weinberg, X. Shuai, M.D. Pagel, D. Farrell, J. Duerk and J. Gao, *Advanced Materials*, 2005, **17**, 1949.
62. G. Kwon, M. Naito, M. Yokoyama, T. Okano, Y. Sakurai and K. Kataoka, *Langmuir*, 1993, **9**, 945.
63. K.N. Prasad, T.T. Luong, A.T. FlorenceJoelle Paris, C. Vaution, M. Seiller and F. Puisieux, *Journal of Colloid and Interface Science*, 1979, **69**, 225.
64. S. Biswas, O.S. Vaze, S. Movassaghian and V.P. Torchilin in *Drug Delivery Strategies for Poorly Water-Soluble Drugs*, John Wiley & Sons Ltd, Hoboken, NJ, USA, 2013, p.411.
65. Y. Lu, and K. Park, *International Journal of Pharmaceutics*, 2013, **453**, 198.
66. S.B. La, T. Okano and K. Kataoka, *Journal of Pharmaceutical Sciences*, 1996, **85**, 85.
67. A.V. Kabanov, E.V. Batrakova and V.Y. Alakhov, *Journal of Controlled Release*, 2002, **82**, 189.
68. A.V. Kabanov and V.Y. Alakhov in *Critical Reviews^{TM} in Therapeutic Drug Carrier Systems*, Ed., M. Sachdeva, Begell House, Inc., Banbury, CT, USA, 2002, p.19.
69. Y. Wang, L. Chen, L. Tan, Q. Zhao, F. Luo, Y. Wei and Z. Qian, *Biomaterials*, 2014, **35**, 6972.
70. K. Letchford, R. Liggins and H. Burt, *Journal of Pharmaceutical Sciences*, 2008, **97**, 1179.
71. J. Lee, S.C. Lee, G. Acharya, C. Chang and K. Park, *Pharmaceutical Research*, 2003, **20**, 1022.
72. K.M. Huh, S.C. Lee, Y.W. Cho, J. Lee, J.H. Jeong and K. Park, *Journal of Controlled Release*, 2005, **101**, 59.
73. M. Morishita and N.A. Peppas, *Drug Discovery Today*, 2006, **11**, 905.
74. V.P. Torchilin, *Pharmaceutical Research*, 2007, **24**, 1.
75. S.K. Shukla, A.W. Shaikh, N. Gunari, A.K. Bajpai and R.A. Kulkarni, *Journal of Applied Polymer Science*, 2009, **111**, 1300.
76. M. Pekař, *Frontiers in Materials*, 2015, **1**, 35.
77. C. Ju, J. Sun, P. Zi, X. Jin and C. Zhang, *Journal of Pharmaceutical Sciences*, 2013, **102**, 2707.
78. L. Bromberg, *Expert Opinion on Drug Delivery*, 2005, **2**, 1003.

79. T.S. Anirudhan, J. Parvathy and A.S. Nair, *Carbohydrate Polymers*, 2016, **136**, 1118.
80. T. Inoue, G. Chen, K. Nakamae and A.S. Hoffman, *Journal of Controlled Release*, 1997, **49**, 167.
81. D. Missirlis, N. Tirelli and J.A. Hubbell, *Langmuir*, 2005, **21**, 2605.
82. S. Merino, C. Martín, K. Kostarelos, M. Prato and E. Vázquez, *ACS Nano*, 2015, **9**, 4686.
83. G.H. Gwak, S.M. Paek and J.M. Oh, *European Journal of Inorganic Chemistry*, 2012, 5269.
84. J. Zhang and P.X. Ma, *Advanced Drug Delivery Reviews*, 2013, **65**, 1215.
85. X. Gao, X. Deng, X. Wei, H. Shi, F. Wang, T. Ye, B. Shao, W. Nie, Y. Li, M. Luo, C. Gong and N. Huang, *International Journal of Nanomedicine*, 2013, **8**, 2453.
86. S.J. Buwalda, K.W.M. Boere, P.J. Dijkstra, J. Feijen, T. Vermonden and W.E. Hennink, *Journal of Controlled Release*, 2014, **190**, 254.
87. P. Matricardi, C. Di Meo, T. Coviello, W.E. Hennink and F. Alhaique, *Advanced Drug Delivery Reviews*, 2013, **65**, 1172.
88. A.S. Hoffman, *Advanced Drug Delivery Reviews*, 2013, **65**, 10.
89. N.A. Peppas and D.S. Van Blarcom, *Journal of Controlled Release*, 2015, **240**, 142.
90. R.K. Verma and S. Garg, *Pharmaceutical Technology*, 2001, **25**, 1.
91. D. Klinger and K. Landfester *Polymer*, 2012, **53**, 5209.
92. R.A. Siegel, *Journal of Controlled Release*, 2014, **190**, 337.
93. Q. Wu, L. Wang, X. Fu, X. Song, Q. Yang and G. Zhang, *Journal of Polymer Research*, 2014, 1.
94. E.S. Gil, and S.H. Hudson, *Progress in Polymer Science*, 2004, **29**, 1173.
95. H. Tian, Z. Tang, X. Zhuang, X. Chen and X. Jing, *Progress in Polymer Science*, 2012, **37**, 237.
96. E. Cabane, X. Zhang, K. Langowska, C.G. Palivan and W. Meier, *Biointerphases*, 2012, **7**, 9.
97. M.A. Ward and T.K. Georgiou, *Polymers*, 2011, **3**, 1215.
98. I.Y. Galaev and B. Mattiasson, *Trends in Biotechnology*, 1999, **17**, 335.
99. Z. Li, S. Cho, I.C. Kwon, M.M. Janát-Amsbury and K.M. Huh, *Carbohydrate Polymers*, 2013, **92**, 2267.
100. D. Chitkara, A. Shikanov, N. Kumar and A.J. Domb, *Macromolecular Bioscience*, 2006, **6**, 977.
101. S. Chen, H. Zhong, B. Gu, Y. Wang, X. Li, Z. Cheng, L. Zhang and C. Yao, *Materials Science and Engineering C*, 2012, **32**, 2199.
102. D. Ko, U. Shinde, B. Yeon and B. Jeong, *Progress in Polymer Science*, 2013, **38**, 672.
103. D. Schmaljohann, *Advanced Drug Delivery Reviews*, 2006, **58**, 1655.
104. E. Fleige, M.A. Quadir and R. Haag, *Advanced Drug Delivery Reviews*, 2012, **64**, 866.
105. A. Imaz, J.I. Miranda, J. Ramos and J. Forcada, *European Polymer Journal*, 2008, **44**, 4002.
106. K. Madhusudana Rao, B. Mallikarjuna, K.S.V. Krishna Rao, S. Siraj, K. Chowdoji Rao and M.C.S. Subha, *Colloids and Surfaces B: Biointerfaces*, 2013, **102**, 891.
107. S. Mendrek, A. Mendrek, H-J. Adler, A. Dworak and D. Kuckling, *Colloid and Polymer Science*, 2010, **288**, 777.
108. D.M. Devine, and C.L. Higginbotham, *European Polymer Journal*, 2005, **41**, 1272.
109. N.V. Gupta and H.G. Shivakumar, *DARU Journal of Pharmaceutical Sciences*, 2010, **18**, 200.
110. G.H. Gao, Y. Li and D.S. Lee, *Journal of Controlled Release*, 2013, **169**, 180.
111. F-J. Xu, E-T. Kang and K-G. Neoh, *Biomaterials*, 2006, **27**, 2787.
112. T. Ishida, Y. Okada, T. Kobayashi and H. Kiwada, *International Journal of Pharmaceutics*, 2006, **309**, 94.
113. N.K. Singh, Q.V. Nguyen, B.S. Kim and D.S. Lee, *Nanoscale*, 2015, **7**, 3043.
114. T.N. Vo, A.K. Ekenseair, F.K. Kasper and A.G. Mikos, *Biomacromolecules*, 2014, **15**, 132.
115. S. Maya, B. Sarmento, A. Nair, N.S. Rejinold, S.V. Nair and R. Jayakumar, *Current Pharmaceutical Design*, 2013, **19**, 7203.
116. J. Gan, X. Guan, J. Zheng, H. Guo, K. Wu, L. Liang and M. Lu, *RSC Advances*, 2016, **6**, 32967.
117. D. Kokuryo, S. Nakashima, F. Ozaki, E. Yuba, K-H. Chuang, S. Aoshima, Y. Ishizaka, T. Saga, K. Kono and I. Aoki, *Nanomedicine: Nanotechnology, Biology and Medicine*, 2015, **11**, 229.

118. N.K. Singh, and D.S. Lee, *Journal of Controlled Release*, 2014, **193**, 214.
119. K. Kono, T. Ozawa, T. Yoshida, F. Ozaki, Y. Ishizaka, K. Maruyama, C. Kojima, A. Harada and S. Aoshima, *Biomaterials*, 2010, **31**, 7096.
120. N.A. Cortez-Lemus and A. Licea-Claverie, *Progress in Polymer Science*, 2015, **53**, 1.
121. M. Prabaharan, J.J. Grailer, D.A. Steeber and S. Gong, *Macromolecular Bioscience*, 2009, **9**, 744.
122. G. Venkatareddy, M. Subha, C. Rao, B. Yerriswamy, L.N. Reddy and V. Prasad, *Asian Journal of Pharmaceutics*, 2010, **4**, 200.
123. M. Prabaharan, J.J. Grailer, D.A. Steeber and S. Gong, *Macromolecular Bioscience*, 2008, **8**, 843.
124. I.S. Cho, C.G. Park, B.K. Huh, M.O. Cho, Z. Khatun, Z. Li, S-W. Kang, Y. Bin Choy and K.M. Huh, *Acta Biomaterialia*, 2016, **39**, 124.

Martin Forde and Ian Major

4 Potential for biodegradable polymer-based nanotechnology in drug delivery

4.1 Introduction

'Nanotechnology' is the ability to design and develop small structures between one to several hundred nanometres of size. To put this in perspective, one nanometre is one-billionth of a metre or one-thousand-times smaller than a micron [1]. There is an enormous capacity for nanostructures to be effective in drug-delivery systems. Several nano-systems have been studied for potential applications in drug and gene delivery [2]. Shrivastava and co-workers highlighted the huge potential in the field of biomedical science where medical nano-devices could one day be routinely injected into patients to monitor and repair conditions that deviate from normal [1]. In 2009, van Boogaard and her team discussed the current problems with co-infection of tuberculosis and the human immunodeficiency virus (HIV). Combined treatment of these diseases involves a high pill burden, as well as side effects from drug–drug interactions between antiretroviral inhibitors and rifampin. The biggest disadvantage of this type of therapy is that drugs are not target-specific [3, 4]. Other limitations of current drug-delivery systems include suboptimal bioavailability and potential cytotoxicity. Current drug-delivery systems are excellent for releasing drugs in a controlled manner to produce a high local concentration. However, this approach is limited to targeting tissues and not individual cells.

Nanotechnology is interdisciplinary, and combines chemistry, biology, engineering and medicine. It has revolutionised the development of drug-delivery systems and devices. Novel biodegradable polymer-based nanotechnology formulations are facilitating site-specific targeting and controlled released of traditional drugs. Polymer nanoscale drug-delivery systems can be created to 'tune' the release kinetics, regulate bio-distribution, and to minimise toxic side effects, thus enhancing the therapeutic index of a given drug [5, 6].

Soumya and Hela (2013) successfully improved drug targeting, specifically in lung cancer, by wrapping pemetrexed in a charged molecule trapped inside a nano-carrier with a particle size of ≈50 nm. The nanoparticle (NP) was composed of biodegradable polyethylene glycol (PEG) in a PEGylating process. PEGylating helped mask the drug in immune systems for longer periods, therefore reducing clearance and optimising its effect. Cancerous cell pores vary from between 10 and 100 nm, and the pores of normal cells are <10 nm, thereby making these drug vehicles highly selective for cancerous cells. Silver nanoparticles (AgNP) were used to tag the over-expressed folate receptor on the cancer cell where curcumin, an antioxidant, acts as a pH indicator and changes colour upon binding. After contact, a current was

https://doi.org/10.1515/9783110640571-004

applied to the solution, where the charged molecule (Prussian blue) lost its negative charge. This initiated the disintegration of the nano-formulation to release the drug at the target site. By turning the voltage on and off it was possible to control the quantity and the timing of the dose [7].

4.2 The bio-nano interface

NP surfaces interact with the biological environment. These interactions are in a dynamic exchange with biomolecules such as proteins, surface receptors and cell organelles (Figure 4.1) [8]. At the nanoscale, this interaction is also called a 'bio-nano interface'. The duration of the interaction between the NP and proteins can affect the way the cells interact with the NP [9]. Proteins that are bound to the NP or are not recognised by the cell will make the NP 'less tasty' to the cell. After all, cells only require what they need. Consequently, these interactions will influence the efficacy of NP. Therefore, a complete understanding is essential when designing NP for therapeutic activity [10]. If the hydrophobic nature of the NP is greater than the cell membrane itself,

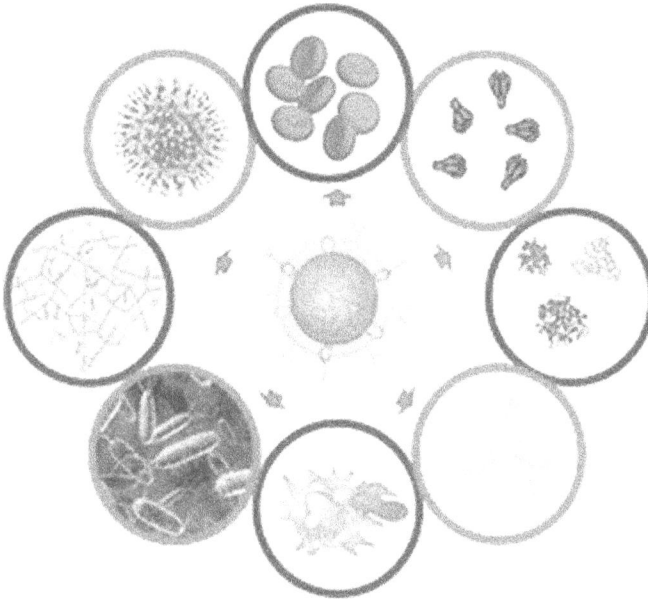

Figure 4.1: Upon contact with biological fluid, NP take on various biological substances that interact with their surroundings. Reproduced with permission from C-M. (Jack) Hu in *Manipulating Nano/Bio Interface*, Research Group of Institute of Biomedical Sciences, Academia Sinica, Taipei, Taiwan, 2015. ©2015, C-M. (Jack) Hu [8].

facilitation to the cells interior may be easier. Thus, the uptake of NP for cell-membrane interaction is very dependent on the many complex characteristics of the NP [10, 11].

4.3 Size and shape matter

To understand the basic concept of nanotechnology, size matters. A 100-nm NP will not inevitably behave in the same way as a 10-nm NP. Below the 100-nm range, natural mechanical effects start to prevail and the properties of materials are affected by particle size. Particles larger than 10 nm are picked up by macrophages and destroyed by the body's immune system. As the applications of polymeric NP advance, it is becoming apparent that shape matters too [12]. A study carried out in 2013 by Kolhar and associates showed that rod-shaped NP exhibit higher specificity compared with their spherical counterparts. Kolhar and his team developed agents to target the diseased endothelium. In an *in vitro* study utilising microfluidic systems that mimic the vasculature, they found that not only did rod shapes have better affinity, but also had lower non-specific accumulation under flow at the target site in comparison with spherical NP [13]. Li and his team designed bovine serum albumin gold nanoparticles (AuNP) of sizes between 40 and 70 nm of spherical and rod-shaped NP for biomedical applications. The coated NP were investigated for their effect on osteogenic differentiation of human mesenchymal stem cells (hMSCs). The spheres of 40 nm, 70 nm, and rods of 70 nm had considerably increased the alkaline phosphatase (ALP) activity and calcium deposition of cells, whereas the rods of 40 nm had reduced ALP activity and reduced calcium depositions. Expression of an osteogenic gene maker exhibited down regulation after incubation with 40 nm of rod-shaped NP. It was concluded that the size and shape of these AuNP had a direct regulatory effect on the osteogenic differentiation of hMSCs [14].

4.3.1 Polymeric nanoparticles

Polymeric NP are usually biocompatible and biodegradable polymers, and vary from natural to synthetic. Their unique properties make them selective for biological distribution where drug absorption is enhanced, therefore increasing the bioavailability [15, 16].

Polymers have a considerable role in metals as a host material and semiconductor NP, where the latter demonstrates excellent optical and electric properties. Nanofillers can be dispersed throughout a polymer where the average particle size is <100 nm [17]. It is evident that drug activity increases when it is conjugated with different polymers [15]. The incorporation of NP with various biodegradable polymers consequently increases drug efficiency [3].

In this chapter, we present various biodegradable polymer NP, and biodegradable polymers combined with noble metal NP such as silver (Ag), gold (Au) or platinum (Pt). These polymer/metal conjugates play a pivotal part in advanced drug-delivery systems.

4.4 Silver nanoparticles conjugated with polymers

Over the past few decades, Ag has been used extensively in human healthcare to treat diseases [3, 18]. The silver ion (Ag$^+$) is biologically active and is readily available to interact with bacteria. Ag ionises in biological fluid and can interrelate with proteins, amino acids and receptors on mammalian cells [18]. Ag has shown potential activity against drug-resistant bacteria, which are a major public health concern [19]. Several studies have been carried out on AgNP as a potential for a novel drug delivery to fight infections and diseases [3]. Numerous studies have investigated the potential effect of AgNP in cancer cell lines, but the effect of AgNP on the cellular mechanism of leukemic cell lines has not been fully elucidated [20]. The unique physiochemical properties of NP, such as particle size, surface area, solubility, and binding affinity to biological sites, all dictate biocompatibility [21, 22]. However, whether AgNP induce cytotoxicity due to Ag ions, AgNP or both is not known [23–26].

AgNP have demonstrated antiviral activity against HIV-1. In 2010, Lara and co-workers carried out a study to determine the mode of antiviral action utilising a panel of *in vitro* assays. The NP were 30–50 nm and were surface-coated with 0.2% biodegradable polyvinylpyrrolidone (PVP). It was concluded that AgNP inhibited viral replication at an early stage of the replication process. It is understood that AgNP binds to glycoprotein 120, where it prevents cluster of differentiation 4-dependent virion binding, fusion and consequently blockade of HIV-1. The PVP-coated AgNP acted as an effective virucide [27]. Lara and co-workers followed up this study by utilising the PVP-coated AgNP-effective properties, and developed a potential topical vaginal microbicide to prevent HIV infection. A concentration of 0.15 mg/ml of PVP-coated AgNP was used on a human cervical culture as an *in vitro* model to replicate *in vivo* conditions. Only a 1-min interval of pre-treatment was needed to prevent HIV-1 transmission to the explant. Furthermore, AgNP were non-toxic to the explant tissue. The data showed the potential for AgNP against HIV infection as a topical agent [28].

In a study carried out in 2013, Guo and associates coated three different particle sizes of AgNP with biodegradable PVP utilising an electrochemical method. Biological studies on cell lines were carried out along with isolates taken from patients with acute myeloid leukaemia (AML). It was concluded that cell viability was reduced through reactive oxygen species (ROS) if Ag^+ was released, inevitably disrupting deoxyribonucleic acid (DNA) and leading to apoptosis. That report highlighted the fact that biodegradable PVP-coated AgNP of various sizes had a direct effect against human AML cell lines and primary isolates from AML patients. Comparatively, the cytotoxic effect was stronger against AML cells than normal haematopoietic cells. The mode of action for the cytotoxic effect was determined using a fluorescence method and compared with a control. It was concluded that Ag^+ release had a significant role in the cytotoxic effect through ROS generation [29].

4.4.1 Silver nanoparticles for gene technology

Gene therapy has been utilised to combat cancer cells. The ultimate goal of gene therapy is to deliver genetic material into new cells without eliciting toxicity on normal tissues. Gene therapy has the potential to inhibit cancer cells in patients with recurrent or metastatic cancer that is often considered incurable. However, progress has been slow due to safety and delivery issues [30].

A one-pot synthesis was carried out by Mukhopadhyay and her team in India using a biodegradable PEG stabilised chitosan (CS)-g-polyacrylamide/AgNP conjugate. CS is a biodegradable, biocompatible, non-toxic natural polymer. Due to the cationic nature of CS, it can encapsulate proteins and drug molecules. CS has shown to extend drug life in the gastrointestinal tract [31]. However, CS is insoluble in water due to its intramolecular and intermolecular hydrogen bonding. Du and Hsieh successfully copolymerised PEG with CS to improve solubility in water and organic solvents [32]. The size of the NP was approximately 38 ± 4 nm. To enhance the efficiency of gene transfection, a tetrapeptide Arg–Gly–Asp–Ser was attached onto AgNP. The NP was stable for ≤30 days after PEG alteration. The transfection efficiency was considerably higher compared with that for polyethylenimine (which has also been used as a delivery agent for gene therapy). This study highlights the potential use of biodegradable polymer-based NP as efficient non-viral carriers for gene therapy [33].

4.4.2 Silver nanoparticles for rheumatoid arthritis

In 2013, Sekar and co-workers showed how AgNP could be used to treat rheumatoid arthritis by controlling the release of the azathioprine and improving its efficiency. Azathioprine is a well- known effective compound for the treatment of rheumatoid arthritis. Utilising a prodrug method, the drug was chemically modified to enhance drug delivery. In their study, Sekar and his team first used CS to stabilise and reduce silver nitrate ($AgNO_3$) to AgNP. Then, AgNP was conjugated to the drug azathioprine. An *in vitro* drug release study and a toxicity test of the AgNP-loaded azathioprine on the 3T3 NIH fibroblast cell line concluded that these biodegradable polymer- based NP produced a synergistic effect to treat rheumatoid arthritis at the active site and release the drug in a controlled manner [34].

4.5 Gold nanoparticles conjugated with polymers

Au is a classic noble metal element. Au is exceedingly unreactive and is evident in artefacts where its gloss can remain pristine for thousands of years [35]. In 1935, Jacques Forestier proved that Au salts can help relieve the symptoms associated with rheumatoid arthritis [36]. Since then, numerous biomedical applications have been studied for the treatment of cancer, HIV and malaria [37]. It is easy to attach molecules to Au; these Au particles can act as a platform, thereby facilitating placement of drugs or proteins within cancerous cells while controlling their release. Decorating AuNP with biodegradable polymers and proteins can improve the stability and target specificity of therapeutic agents [38].

A novel, versatile, AuNP drug-delivery system was developed by Wang and co-workers based on an amphiphilic diblock copolymer: PEG-*b*-poly(n-butyl acrylate). The copolymer was synthesised by reversible addition-fragmentation transfer (RAFT) polymerisation whereby a thiocarbonylthio group at the end of the hydrophobic block was reduced to a thiol group. RAFT polymerisation is applicable to a wide range of vinyl polymers, and is known to be versatile without the need for protecting groups. The polycyclic hydrocarbon model drug pyrene was entrapped by hydrophobic interactions in the shell of AuNP. Ultraviolet analyses of an AuNP- pyrene composite showed good stability in aqueous solution and high-drug load capacity compared with AuNP decorated with smaller ligands. In addition, the study concluded that the chain length of the hydrophobic block determined the loading capacity and that this delivery system may be useful to produce efficient AuNP drug carriers [39].

5-Fluorouracil (5-FU) is an anticancer drug and has been used for decades. The drug has been given systemically for breast, stomach, anal and particularly head and neck skin cancers. Over the years understanding of its mode of action has led researchers to enhance its anticancer activity. Most anticancer drugs attack normal cells and cancer cells, leading to toxic side effects. To overcome this problem, Sathishkumar (2012) designed a biodegradable AuNP to control the rate of 5-FU and to improve site specificity. Sathishkumar first synthesised the copolyester polylactic acid (PLA)-*co*-ethyl cellulose (EC) by azeotropic dehydration of the lactic acid and EC. 5-FU was encapsulated by the prepared polymer in the presence and absence of AuNP by solvent evaporation. The copolymer in the presence of AuNP indicated a slower, controlled drug release compared with 5-FU-incorporated PLA-*co*-EC nanocapsules. This data suggests enhanced interaction between the drug and polymer-based AuNP [40].

Utilising the selectivity of cancerous cells has been well documented. Huang, El-Sayed and their team demonstrated the use of tumour selectivity in cancer cells by conjugating epithelial growth factor receptor (EGFR) to a AuNP and nanorods. In both studies the results showed the potential use of this selective method. The EGFR antibody that was conjugated to nanorods/ NP bound specifically to the surface of cancer cells with much higher affinity than that observed with a control [41, 42]. The pharmacologically active agent tamoxifen has been administered extensively for the treatment of breast cancer for more than three decades. The drug has been known to compete with 17β-estradiol for binding affinity to an oestrogen receptor that is upregulated in breast cancer cells to initiate cell death. In 2010, Dreaden and co-workers synthesised a biodegradable thiol-PEGylated tamoxifen derivative AuNP that was capable of a 2.7-fold enhanced potency in cancer cell lines *in vitro*. Uptake of the polymer NP/drug conjugate was compared with that of the free drug utilising a time-dependent dose-response study. The AuNP conjugate moved into cells by endocytosis and free drug by passive diffusion. There were no adverse effects associated with the AuNP themselves, and the evidence suggested that the plasma membrane-localised α-estradiol could facilitate endocytotic transport of these NP conjugates. Tamoxifen could, therefore, not only act as a selective targeting agent, but also as a powerful treatment for malignancies which overexpress oestrogen [43].

4.6 Platinum nanoparticles conjugated with polymers

In 1965, Barnett Rosenberg carried out an investigation of the effects of an electric field on the growth process on the bacteria *Escherichia coli* by exploiting a Pt-based compound. It was concluded that the metal in the culture media inhibited the growth of Gram-negative bacteria [44]. Rosenberg and his team followed up this study in 1969, whereby they discovered that various Pt compounds, predominately cisplatin, could be vastly effective against sarcoma 180 and leukaemia L1210 in mice [45]. In 1971, the drug was approved by the US Food and Drug Administration (FDA), and entered clinical trials. Many Pt drugs and their structure–activity relationship have provided important insights into how to develop the clinical drug cisplatin. Nonetheless, there are still many problems associated with cisplatin, including drug resistance and the side effects it produces [46]. Cisplatin is not a NP; however, platinum nanoparticles (PtNP) have been shown to retain the ability to enter host cells [47].

In 2010, Kolishettia and his co-workers designed a self-assembled polymeric biodegradable polymer NP. Two anticancer drugs, cisplatin and docetaxel, were controlled precisely and could be delivered to prostate cancer cell lines with synergistic cytotoxicity. A polylactide derivative, decorated with hydroxyl groups, was synthesised and conjugated with Pt(IV) whereby the hydroxyl group was reduced, resulting in a hydrophilic prodrug. A dual-drug delivery NP was assembled by combining the carboxyl-terminated poly(lactic-*co*-glycolic acid) (PLGA)-block-PEG copolymer in the presence and absence of docetaxel. This nanoprecipitation step entrapped the hydrophobic docetaxel, leaving the affixed hydrophilic Pt(IV) on the outside. The NP surface was coded with an A10 aptamer, which bound to prostate-specific membrane-expressing cancer cells. Aptamers are used specifically as binding proteins and are quite selective. Targeted dual-combinational NP with docetaxel and Pt(IV) showed exceptional efficacy over a single drug NP *in vitro*. These complex dual polymer-NP systems expose the potential to treat diseases, where combination therapy is desired [48].

A similar method was employed by Xiao and associates for delivering the anticancer drug cisplatin intravenously and paclitaxel (PTX). The polymer–cisplatin IV conjugate was developed by a tri-block amphiphilic biodegradable PEG-*b*-PCL-*b*-poly (L-lysine). This system was not however suitable for PTX conjugation. A separate biodegradable, biocompatible diblock copolymer, PEG-*b*-PLA-*co*-microcrystalline cellulose, was used as the carrier system for PTX. The products were co-assembled in a one-step reaction. *In vitro* studies showed the polymer-based combination of the nano-micelles was more safe and efficacious against U14 cancer cells compared with the free drug. Furthermore, the complex displayed synergistic effects: cisplatin was released upon cellular reduction and PTX by acidic hydrolysis after entering cancerous cells. This dual PtNP combination could pave the way to co-deliver drugs at the nanoscale [49].

In 2014, Babu and his team designed multi-PtNP to overcome resistance and restore sensitivity to anticancer drugs. Several studies have shown that the proteasome and regulatory protein P62 contribute to cisplatin resistance. Reduction in the expression of the $\beta5$ subunit of the proteasome and enhanced expression of P62 protein is known to specifically contribute to drug resistance (cisplatin) in ovarian cancer cells. Babu and co-workers hypothesised that suppression of P62 protein expression and restoring $\beta5$ expression in ovarian cancer cells would help elucidation of the anti-tumour activity of cisplatin. Babu and co-workers tested their hypothesis by developing a biodegradable multifunctional nanoparticle (MNP) that co-delivered P62 small-interfering ribonucleic acid (siRNA), $\beta5$ plasmid deoxyribonucleic acid (pDNA) and cisplatin, and used it against ovarian cancer cells. The MNP core consisted of cisplatin in PLA, and a cationic CS that was linked with P62 ribonucleic acid (siRNA) and $\beta5$-expressing pDNA as the outer shell. MNP efficiently protected siRNA and showed promising stability compared with naked siRNA. The MNP effectively delivered siP62 and p$\beta5$ to cause P62 knockdown while restoring $\beta5$ expression in cells. This combined delivery system, utilising biodegradable MNP, resulted in a noticeable reduction in the half-maximal inhibitory concentration (IC_{50}) in cells compared with the reduction of IC_{50} in cells in which only siP62 was delivered.

Additionally, the mean number of resistant cisplatin cells was reduced from 4.62 for free cisplatin to 3.62 for MNP treatment. Therefore, these results implied the efficacy of the polymeric MNP system in overcoming cisplatin resistance in ovarian cancer cells [50].

4.7 Other polymer non-metal nanoparticles for smart drug delivery

Systemically administrated chemotherapy drugs have been ineffectual for the treatment of primary and metastatic brain tumours. The blood–brain barrier (BBB) plays a pivotal part in preventing drug delivery to the active-tumour site. The BBB is considered a key attribute for the lack of efficacy and low penetration of chemotherapy drugs [51]. The anticancer drug PTX has not been fully elucidated for the treatment of brain cancers because it cannot cross the BBB due to efflux by P-glycoprotein (P-gp). Recently, Geldenhuys and his associates in Ohio coated biodegradable PLGA-PEG NP with glutathione with the intent of helping PTX-loaded NP across the BBB. PTX-loaded NP have shown efficacy *in vivo* in brain tumour-bearing mouse models. *In vitro* analysis confirmed there was no burst release evident by the NP and showed sustained drug release. The particle size of the NP for BBB permeation showed a higher cellular uptake in rat glioma 2 (RG2) cells compared with uncoated NP. The coated NP showed considerably higher toxicity in RG2 cells compared with uncoated NP or free PTX in solution. Geldenhuys and co-workers utilised tubulin immunofluorescence to show higher cell death by coated NP due to increased microtubule stabilisation. These glutathione-coated biodegradable polymer-based NP could be used for effective delivery of PTX for the treatment of brain cancers [52].

In 2012, Bahadur and co-workers designed a novel eptide cyclic (Arg–Asp–D–Phe–Cys) (RGD) with a biodegradable poly[2-(pyridine-2-yldisulfanyl) ethyl acrylate]-g-PEG. The side chain 2-(pyridine-2-yldisulfanyl) was synthesised by free-radical polymerisation and was conjugated to the copolymer/peptide through a thiol-disulfide exchange reaction. The anticancer drug doxorubicin was entrapped into the NP using the co-solvent dialysis method. The NP conjugate showed excellent stability in physiological conditions and the drug was released quickly by activation of acidic pH and redox potential. Additionally, the NP system showed two-phase release kinetics. A fast-release phase provided a minimal-effective concentration of the drug to inhibit cell growth, and a slow release phase that offered a maintenance dose for cancer therapy. The RGD peptide, amalgamated to the NP, enhanced cellular uptake and nuclear confinement of the nano-system. Compared with free doxorubicin, the NP conjugate exhibited similar IC_{50} results for HCT-116 colon cancer cells.

Moreover, the NP conjugate displayed notably higher anticancer efficacy than the free drug, due mainly to the synergistic effect of RGD targeting and two-phase kinetic release. The results demonstrate the potential of tumour-targeted chemotherapy by nanotechnology [53].

Resveratrol (RSV) is a compound found in grapes, berries, red wine and other foods. It has shown promising anticancer prevention by inhibiting tumour initiation, promotion and progression [54–56]. Encapsulating RSV through polymeric NP is likely to protect the natural compound against degradation and enhance its bioavailability, while improving intracellular penetration and controlling delivery. In 2013, Sanna and co-workers encapsulated polyphenol RSV into a biocompatible blend of PCL and PLGA–PEG-COOH conjugated by a nanoprecipitation method. Due to a high RSV-loading capacity, and controlling the release rate at different pH values, the drug could by administered *via* oral or parenteral routes. NP uptake was evaluated in prostate cancer cells, where anti-proliferation efficacy was superior compared with that of free RSV. This enhanced uptake was attributed to endocytosis. These small biodegradable polymeric NP have the potential to deliver bioactive RSV for the treatment of prostate cancer [54].

4.8 Biomedical implants

One of the most common cardiovascular diseases is atherosclerosis, a build-up of plaque in arteries. This build up can accumulate to where a clot can form, blocking the artery completely and leading to a stroke or heart attack. If atherosclerosis occurs in the heart, it is known as coronary artery disease. Cardiovascular diseases are a major health concern worldwide [57]. The initial treatment of atherosclerosis used to be coronary artery bypass grafting (CABG). CABG is a surgical procedure where arterial and saphenous vein grafts are used to bypass the obstructed path. Percutaneous transluminal coronary angioplasty has replaced CAGB and is less invasive. A small incision is made in the groin under local anaesthesia. A catheter with a balloon attached can be fed through the artery and inflated to unblock the artery [58]. However, there are two major limitations of balloon angioplasty: incomplete revascularisation and restenosis (re-narrowing of the blood vessel) [59]. In 1994, the FDA approved the use of stents because they appeared to have an average 10% lower rate of restenosis compared with balloon angioplasty [60]. Nevertheless, this figure appeared to be higher in follow-up studies conducted in the same year [61, 62]. To overcome this problem, drug-eluting stents were utilised to decrease restenosis and minimise revascularisation. Conversely, a report published in 2007 claimed there was an increase in late stent thrombosis with drug-eluting stents compared with bare-metal stents [63, 64]. Recent studies have been carried out on cardiovascular biomedical implants aimed at improving efficacy and safety [63].

A sirolimus-loaded biodegradable poly(D,L-lactic acid) (PDLLA) NP was developed in 2011 to prevent restenosis after stent implantation. Sirolimus has known potent immunosuppressive and anti-inflammatory properties and been shown to decrease in-stent restenosis in animal studies. Luderer and his team incorporated 20% sirolimus in PDLLA and carried out proliferation and cytotoxicity studies in human coronary artery smooth muscle cells (HCASMC) and human coronary artery endothelial cells (HCAEC). The sirolimus-loaded NP inhibited the viability and proliferation of HCASMC and HCAEC. The drug-loaded PDLLA was less effective on endothelial cells compared with the free drug. Consequently, the sirolimus-loaded PDLLA NP could prevent restenosis by inhibition of smooth muscle cell proliferation without inhibiting endothelial cells and, thus, may represent a promising therapeutic approach for angioplasty and stent technology [65].

In 2013, Haeri and co-workers designed a biocompatible, biodegradable nanomicelle from PEG conjugates with phosphatidylethanolamine and PEGylated liposomes loaded with sirolimus. The sirolimus-loaded micelles (14 nm) and liposomes (90 nm) were analysed for their anti-restenotic effects in a rat model of carotid injury. The drug-loaded micelles and nanoliposomes significantly reduced vascular stenosis by 42 and 19%, respectively, compared with control groups. The results imply that the biodegradable phospholipid-based micelles provide better anti-restenotic effects than the PEGylated liposomes due to the smaller particle size of the phospholipid-based micelles. The results show that both biodegradable colloidal nanocarriers could be used effectively for drug delivery and treatment of restenosis, providing a more efficacious and safer alternative to conventional methods [66].

4.8.1 Neuronal implants

Epilepsy is a serious neurological condition characterised by impulsive seizures. More than 30% of patients are in treatment with oral anti-epileptic drugs (AED), which are considered insufficient. Scholars are searching for alternatives to deliver AED at the active site without interfering with other parts of the brain and organs and to prevent side effects [67]. β-Carotene (at 200 mg/kg body weight) has been well documented for the treatment of epileptic seizures. In a recent study, β-carotene-containing polysorbate 80-coated with biodegradable PLGA NP was prepared by a solvent diffusion method. The polymer vehicle delivered a 2 mg/kg dose through the intraperitoneal route in epileptic albino mice and was compared with an unformulated β-carotene dose, and was shown to be effective. This novel coated NP was stable, delivered a sustained release where the desired bioavailability was reached, and thus overcame the problems associated with conventional delivery systems [68].

Thyrotropin-releasing factor (THF) is a hormone known to have anticonvulsant effects in animal models of seizure and unmanageable epileptic patients. However, rapid-tissue metabolism and the BBB limit its duration of action. Kubek and his team developed a THF-loaded biodegradable PLA microdisc NP, and implanted the sustained-release system into the nasal cavity of a rat under aesthesia. Additionally, the NP contained a fluorescent dye which highlighted good NP uptake and distribution *in vivo* and *in vitro*, respectively. This result suggests that neuropeptides and other drugs can be administrated safely and effectively utilising intranasal delivery to suppress epileptic seizures [69].

4.8.2 Cochlear and neural implants

Hearing loss affects more than 44 million people in the European Union alone. Current strategies to treat inner-ear diseases are limited. The cochlea-blood barrier restricts access to the cochlea, making drug delivery by conventional methods difficult. A human nerve growth factor (hNGF)-derived peptide (hhNGF_EE) functionalised to biodegradable polymeric (PEG-*b*-PCL) to treat target cells on the inner ear was developed. The NP conjugates were introduced to organotypic explant cultures of mouse inner ear and PC-12 rat pheochromocytoma cells. There was no evidence of toxicity. Specific targeting and binding to explant cultures was achieved through ligand-mediated multivalent binding to tyrosine kinase receptors and to p75 neurotrophic receptors. Compared with an un-functionalised NP with no peptide hNGF_EE there was an increase in specificity towards spiral ganglion neurons. This study shows the potential to treat inner-ear disease utilising cell specificity for drug and gene delivery [70].

Implanting microelectrode arrays have the potential to improve the quality of life of patients suffering from full or partial paralysis [71]. A microelectrode component was developed by Sun and co-workers in 2014 to record brain activity. A biocompatible porous silicone (Psi) layer was prepared by an anodisation process, followed by incubating the microelectrode in a solution of the anti-inflammatory drug dexamethasone (Dex). The drug-loading pore size was reduced to 4.7 ± 2.6 nm and drug release of Dex-loaded Psi was initially a burst release and subsequently sustained release. Primary astrocyte cell cultures were seeded on the surface of samples to evaluate cellular response. The Dex-loaded Psi had more astrocytes attached to it than unloaded Psi, thus effectively suppressing the activation of primary astrocytes due to drug release. This might be an effective approach to decrease the host-tissue response and alleviate the quality of the neural interface using Dex-loaded Psi [72].

4.9 Toxicity of nanoparticles – a major concern

AgNP have made a remarkable return as potential antimicrobial agents. Ag, as discussed above, combined with various biodegradable polymers, has been well studied for treating various types of bacteria that have become resistant to other forms of antibiotics [73]. However, their mode of antimicrobial activity and toxicity still remain unclear and uncertain [3]. In 2014, Ahn and co-workers carried out toxicity studies on uncoated AgNP, AgNO$_3$, and PVP-coated AgNP that were 8 nm and 38 nm in size and tested against the nematode *Caenorhabditis elegans*. The study claimed that AgNO$_3$ and uncoated AgNP had similar toxicities compared with PVP-coated AgNP, where toxicity was reduced significantly. The 8 nm-coated AgNP had higher toxicity than that of the larger size. AgNO$_3$ and bare AgNP-induced damage to mitochondrial membranes. The authors claimed that polymer-coated AgNP and size have a direct effect on toxicity due to oxidative stress on mitochondria and DNA damage to cell lines [74].

Little is known of the water chemistry and stability of AgNP in the environment. In a study carried out in 2015, Oliver and her team tested the effects of water hardness and humic acid on the bioaccumulation and toxicity of PVP-coated AgNP to the fresh-water snail *Lymnaea stagnalis* after dietary exposure. They concluded that the toxicity and bioaccumulation of Ag from PVP-coated AgNP that was ingested with food were not affected by water hardness and humic acid. However, both could affect interactions with biological membranes and initiate NP transformation [75].

As mentioned above, polymer-coated biodegradable AuNP are being used extensively in biomedical applications. In an interesting study, Connor and associates reported the absence of toxicity with NP with various surface modifiers. The citrate-capped AuNP were 18 nm in average and were not toxic to leukaemia cell lines. The authors argued that, although some precursors of NP may be toxic, the NP themselves may not be necessarily detrimental to cellular function [76]. Additionally, in a study carried out by Lasagna-Reeves and co-workers, the bioaccumulation and toxic effects of different doses of AuNP that contained 40, 200 and 400 µg/kg/day was investigated. The NP had a size of 12.5 nm and were administered to mice through an intraperitoneal injection every day for 8 days. The authors found that the Au levels in blood did not increase with the dose administered. This research team also concluded that the brain contained the lowest amount of AuNP compared with other organs, but that these NP could pass the BBB. In addition, there was evidence of toxicity [77].

On the contrary, the toxicity of three AuNP capped with polyallylamine hydro-chloride (PAH), citrate and mercaptopropionic acid were evaluated on the model species *Daphnia magna*. Acute exposure toxicity assays showed that the overall negatively-charged AuNP were less toxic that the positively-charged AuNP. The study concluded that although PAH indicates minimally toxicity, when conjugated to NP, toxicity is enhanced. Surface chemistry plays an essential part in NP toxicity and capping has the potential to affect the sustainability of these materials [78].

In 2014, chicken embryo red blood cells (RBC) were treated with PtNP to investigate structural damage, cell-membrane deformation, and haemolysis. When the chicken embryo RBC were incubated with 2.6 µg/ml of PtNP, it was detrimental to cell structure and induced haemolysis. Haemolytic injury was increased compared with that in the control group. PtNP have great potential in anticancer therapy but they can cause cell-membrane deformations and formation of knizocytes and echinocytes, leading to toxic side effects [79]. The human skin is constantly exposed to toxic particles. Konieczny and associates initiated a study to address whether biodegradable PVP-coated PtNP have a negative effect on skin cells (mainly epidermal keratinocytes). PtNP of 5.8 and 57 nm were used in the study at concentrations of 6.25, 12.5 and 25 µg/ml. Toxic effects on primary keratinocytes were found. Furthermore, smaller NP were more toxic than larger ones, in which DNA stability was decreased [80].

Biodegradable polybutylcyanoacrylate (PBCA) NP are considered excellent drug carriers that can cross the BBB. In 2014, Voigt and her team repeatedly injected PBCA NP rhodamine-labelled variations into rats and monitored survival for 5 weeks: toxic side effects were not observed. *In vitro* analysis was also done on HeLa and human embryonic kidney 293 cells, in which dose-dependent cell death occurred but only at high doses. These results indicated that these polymeric NP could have the potential to be used as drug-delivery systems with little or no toxicity [81]. On the contrary, positive- and negatively-charged biodegradable PLGA NP of 200 nm were tested for toxicity on human THP-1 macrophages. The NP were prepared utilising poloxamer 188 and PVA as stabilisers and compared with stabiliser-free PLGA NP. At low concentrations, no toxicity was apparent. At high concentrations (>1 mg/ml), cytotoxicity was found to be induced by the presence of stabilisers, whereas the stabiliser-free PLGA NP exerted no cytotoxicity. The results suggested the significant toxicological contribution that stabilisers have for the formulation of PLGA NP, and the implications for local and cellular toxic effects [82].

4.10 Summary

Although considerable research has been carried out in nanotechnology, many areas still require adequate attention. Various metals and biodegradable polymers have attracted great interest in biomedical applications. However, more originality is required to tailor precise surface modifications of these metal-decorated polymer conjugates. There have been many applications of biodegradable polymeric NP systems for the treatment of cardiovascular diseases and in neuronal implants. Nanotechnology has many advantages over traditional methods. However, only clinical studies will reveal if these biodegradable polymeric NP are superior to current strategies. From the literature reviewed, the toxicity of biodegradable polymeric NP systems has scarcely been studied. The mode of action of nano-toxicity is yet to be fully elucidated. Consequently, there is a need for additional research to overcome perilous side effects associated with NP toxicity.

References

1. S. Shrivastava and D. Dash, *Journal of Nanotechnology*, 2009, **2009**, 1.
2. A. Brandelli, *Mini-Reviews in Medicinal Chemistry*, 2012,**12**, 731.
3. M. Rai, A.P. Ingle, I. Gupta and A. Brandelli, *International Journal of Pharmaceutics*, 2015, **496**, 159.
4. J. Den van Boogaard, G.S. Kibiki, E.R. Kisanga, M.J. Boeree and R.E. Aarnoutse, *Antimicrobial Agents and Chemotherapy*, 2009, **53**, 849.
5. M. Goldberg, R. Langer and X. Jia, *Journal of Biomaterials Science: Polymer Edition*, 2007, **18**, 241.
6. O. Kayser, A. Lemke and N. Hernández-Trejo, *Current Pharmaceutical Biotechnology*, 2005, **6**, 3.
7. R.S. Soumya and P.G. Hela, *Scholars Research Library*, 2013,**5**, 189.
8. C-M. (Jack) Hu in *Manipulating Nano/Bio Interface*, Research Group of Institute of Biomedical Sciences, Academia Sinica, Taipei, Taiwan, 2015.
9. I. Lynch, A. Salvati and K.A. Dawson, *Nature Nanotechnology*, 2009, **4**, 546.
10. X. Zhang, *Cell Biochemistry and Biophysics*, 2015, **72**, 771.
11. A.E. Nel, L. Mädler, D. Velegol, T. Xia, E.M.V. Hoek,P. Somasundaran, F. Klaessig, V. Castranova andM. Thompson, *Nature Materials*, 2009, **8**, 543.
12. R. Brazil, *Chemistry World*, 2016, **13**, 65.
13. P. Kolhar, A.C. Anselmo, V. Gupta, K. Pant, B. Prabhakarpandian, E. Ruoslahti and S. Mitragotri, *Proceedings of the National Academy of Sciences*, 2013, **110**, 10753.
14. H. Wang, Y. Liu, M. Li, H. Huang, H.M. Xu, R.J. Hong and H. Shen, *Optoelectronics and Advanced Materials, Rapid Communications*, 2010, **4**, 1166.
15. A.Z. Mirza and F.A. Siddiqui, *International Nano Letters*, 2014, **4**, 94.
16. S. Gelperina, K. Kisich, M.D. Iseman and L. Heifets, *American Journal of Respiratory and Critical Care Medicine*, 2005, **172**, 1487.
17. H. Huang, Q. Yuan and X. Yang, *Colloids and Surfaces B: Biointerfaces*, 2004, **39**, 31.
18. A. Lansdown, *Current Problems in Dermatology*, 2006, **33**, 17.
19. J. Chastre, *Clinical Microbiology and Infection*, 2008, **14**, 3.
20. D. Guo, Y. Zhao, Y. Zhang, Q. Wang, Z. Huang, Q. Ding, Z. Guo, X. Zhou, L. Zhu and N. Gu, *Journal of Biomedical Nanotechnology*, 2014, **10**, 669.
21. C. Carlson, S.M. Hussein, A.M. Schrand, L.K. Braydich- Stolle, K.L. Hess, R.L. Jones and J.J. Schlager, *Journal of Physical Chemistry B*, 2008, **112**, 13608.
22. D-H. Lim, J. Jang, S. Kim, T. Kang, K. Lee and I-H. Choi, *Biomaterials*, 2012, **33**, 4690.
23. X. Yang, A.P. Gondikas, S.M. Marinakos, M. Auffan, J. Liu, H. Hsu-Kim and J.N. Meyer, *Environmental Science and Technology*, 2012, **46**, 1119.
24. C. Beer, R. Foldbjerg, Y. Hayashi, D.S. Sutherland and H. Autrup, *Toxicology Letters*, 2012, **208**, 286.
25. Z. Wang, S. Liu, J. Ma, G. Qu, X. Wang, S. Yu, J. He, J. Liu,T. Xia and G. Bin Jiang, *American Chemical Society Nano*, 2013, **7**, 4171.
26. R.P. Singh and P. Ramarao, *Toxicology Letters*, 2012, **213**, 249.
27. H.H. Lara, N. V Ayala-Nuñez, L. Ixtepan-Turrent and C. Rodriguez-Padilla, *Journal of Nanobiotechnology*, 2010, **8**, 1.
28. H.H. Lara, L. Ixtepan-Turrent, E.N. Garza-Treviño and C. Rodriguez-Padilla, *Journal of Nanobiotechnology*, 2010, **8**, 1.
29. D. Guo, L. Zhu, Z. Huang, H. Zhou, Y. Ge, W. Ma,J. Wu, X. Zhang, X. Zhou, Y. Zhang, Y. Zhao and N. Gu, *Biomaterials*, 2013, **34**, 7884.
30. S. Mali, *Indian Journal of Human Genetics*, 2013, **19**, 3.

31. P. Mukhopadhyay, K. Sarkar, M. Chakraborty, S. Bhattacharya, R. Mishra and P.P. Kundu, *Materials Science and Engineering C*, 2013, **33**, 376.
32. J. Du and Y. Lo Hsieh, *Cellulose*, 2007, **14**, 543.
33. K.C.K. Sarkar, S.L. Banerjee, P.P. Kundu and G. Madrasa, *Journal of Materials Chemistry B*, 2015, **3**, 5266.
34. R.P. Sekar, E. Kulandaivel, D. Damayanthi and J.S. Saranya, *Asian Journal of Biomedical and Pharmaceutical Sciences*, 2013, **3**, 28.
35. E.C. Dreaden, A.M. Alkilany, X. Huang, C.J. Murphy and M.A. El-Sayed, *Chemical Society Reviews*, 2012, **41**, 2740.
36. A. Rachid, *Revista Medica de Panama*, 1934, **19**, 20.
37. C.F. Shaw, *American Chemical Society*, 1999, **99**, 2589.
38. K. Weintraub, *Nature*, 2013, **495**, S14.
39. Z. Wang, L. Jia and M-H. Li, *Journal of Bomedical Nanotechnology*, 2013, **9**, 61.
40. K. Sathishkumar, *International Journal of Nano and Biomaterials*, 2012, **4**, 12.
41. I.H. El-Sayed, X. Huang and M.A. El-Sayed, *Cancer Letters*, 2006, **239**, 129.
42. X. Huang, I.H. El-Sayed, W. Qian and M.A. El-Sayed, *Journal of the American Chemical Society*, 2006, **128**, 2115.
43. A.E.C. Dreaden, S.C. Mwakwari, Q.H. Sodji, A.K. Oyelere and M.A. El-Sayed, *Bioconjugate Chemistry*, 2010, **20**, 2247.
44. B. Rosenberg, L. Vancamp and T. Krigas, *Nature*, 1965, **205**, 698.
45. B. Rosenberg, L. van Camp, J.E. Trosko and V.H. Mansour, *Nature*, 1969, **222**, 385.
46. A.G. Quiroga, *Current Topics in Medicinal Chemistry*, 2011, **11**, 2613.
47. M. Yamada, M. Foote and T.W. Prow, *Wiley Interdisciplinary Reviews: Nanomedicine and Nanobiotechnology*, 2015, **7**, 428.
48. N. Kolishetti, S. Dhar, P.M. Valencia, L.Q. Lin, R. Karnik, S.J. Lippard, R. Langer and O.C. Farokhzad, *Proceedings of the National Academy of Sciences of the United States of America*, 2010, **107**, 17939.
49. H. Xiao, H. Song, Q. Yang, H. Cai, R. Qi, L. Yan, S. Liu, Y. Zheng, Y. Huang, T. Liu and X. Jing, *Biomaterials*, 2012, **33**, 6507.
50. A. Babu, Q. Wang, R. Muralidharan, M. Shanker, A. Munshi and R. Ramesh, *Molecular Pharmaceutics*, 2014, **11**, 2720.
51. E.M. Kemper, W. Boogerd, I. Thuis, J.H. Beijnen and O. van Tellingen, *Cancer Treatment Reviews*, 2004, **30**, 415.
52. W. Geldenhuys, T. Mbimba, T. Bui, K. Harrison and V. Sutariya, *Journal of Drug Targeting*, 2011, **19**, 837.
53. B. Remant, B. Thapa and P. Xu, *Molecular Pharmaceutics*, 2012, **9**, 2719.
54. V. Sanna, I.A. Siddiqui, M. Sechi and H. Mukhtar, *Molecular Pharmaceutics*, 2013, **10**, 3871.
55. Q-B. She, A.M. Bode, W-Y. Ma, N-Y. Chen and Z. Dong, *Cancer Research*, 2001, **61**, 1604.
56. O.P. Mgbonyebi, J. Russo and I.H. Russo, *International Journal of Oncology*, 1998, **12**, 865.
57. D. Kelley, *Johnson County Community College*, 2014, **5**, 1.
58. A. Bakhai, R.A. Hill, Y. Dundar, R.C. Dickson and T. Walley, *Cochrane Library*, 2005, **1**, 1.
59. E.D. Grech, *British Medical Journal (Clinical Research Edition)*, 2003, **326**, 1137.
60. A.K. Mitra and D.K. Agrawal, *Journal of Clinical Pathology*, 2006, **59**, 232.
61. P.W. Serruys, P.D. Jaegere, F. Kiemeneji, C. Macaya, W. Rutsch, G. Heyndrickx, H. Emanuelsson, J. Marco, X. Legrand, P. Materne, J. Belardi, U. Sigwart, A. Colombo, J.J. Goy, P.V.D. Heuvel, J. Delcan and A.M. Morel, *The New England Journal of Medicine*, 1994, **331**, 481.
62. H. Fenton, D.L. Fischman, M.P. Savage, R.A. Schatz, M.B. Leon, D.S. Baim, S.I. King, R.R. Heuser, R.C. Curry, R.C. Rake and S. Goldberg, *American Journal of Cadiologly*, 1994, **74**, 1187.

63. G.G. Stefanini and D.R Holmes, Jr., *New England Journal of Medicine*, 2013, **368**, 3, 254.
64. U. Stenestrand, J. Lindbäck and T. Nilsson, *New England Journal of Medicine*, 2007, **356**, 1009.
65. F. Luderer, M. Löbler, H.W. Rohm, C. Gocke, K. Kunna,K. Köck, H.K. Kroemer, W. Weitschies, K-P. Schmitz and K. Sternberg, *Journal of Bomaterials Applications*, 2011, **25**, 851.
66. A. Haeri, S. Sadeghian, S. Rabbani, M.S. Anvari, A. Lavasanifar, M. Amini and S. Dadashzadeh, *International Journal of Pharmaceutics*, 2013, **455**, 320.
67. A.J. Halliday, S.E. Moulton, G.G. Wallace and M.J. Cook, *Advanced Drug Delivery Reviews*, 2012, **64**, 953.
68. M. Yusuf, R.A. Khan, M. Khan and B. Ahmed, *International Journal of Nanomedicine*, 2012, **7**, 4311.
69. M.J. Kubek, A.J. Domb and M.C. Veronesi, *Neurotherapeutics*, 2009, **6**, 359.
70. S. Roy, A.H. Johnston, T.A. Newman, R. Glueckert, J. Dudas, M. Bitsche, E. Corbacella, G. Rieger, A. Martini and A. Schrott-Fischer, *International Journal of Pharmaceutics*, 2010, **390**, 214.
71. V.S. Polikov, P.A. Tresco and W.M. Reichert, *Journal of Neuroscience Methods*, 2005, **148**, 1.
72. T. Sun, W.M. Tsang and W.T. Park, *Applied Surface Science*, 2014, **292**, 843.
73. M. Rai, A. Yadav and A. Gade, *Biotechnology Advances*, 2009, **27**, 76.
74. J.M. Ahn, H.J. Eom, X. Yang, J.N. Meyer and J. Choi, *Chemosphere*, 2014, **108**, 343.
75. A.L.S. Oliver, M.N. Croteau, T.L. Stoiber, M. Tejamaya, I. Römer, J.R. Lead and S.N. Luoma, *Environmental Pollution*, 2014, **189**, 87.
76. E.E. Connor, J. Mwamuka, A. Gole, C.J. Murphy and M.D. Wyatt, *Small*, 2005, **1**, 325.
77. C. Lasagna-Reeves, D. Gonzalez-Romero, M.A. Barria,I. Olmedo, A. Clos, V.M. Sadagopa Ramanujam,A. Urayama, L. Vergara, M.J. Kogan and C. Soto, *Biochemical and Biophysical Research Communications*, 2010, **393**, 649.
78. J. Bozich, S. Lohse and M. Torelli, *Environmental Science Nano*, 2014, **1**, 260.
79. M. Kutwin, E. Sawosz, J. Jaworski, N. Kurantowicz, B. Strojny and A. Chwalibog, *Nanoscale Research Letters*, 2014, **9**, 257.
80. P. Konieczny, A.G. Goralczyk, R. Szmyd, L. Skalniak, J. Koziel, F.L. Filon, M. Crosera, A. Cierniak, E.K. Zuba-Surma, J. Borowczyk, J. Laczna, J. Drukala, E. Pyza, D. Semik, O. Woznicka, A. Klein and J. Jura, *International Journal of Nanomedicine*, 2013, **8**, 3963.
81. N. Voigt, P. Henrich-Noack, S. Kockentiedt, W. Hintz,J. Tomas and B.A. Sabel, *Journal of Nanoparticle Research*, 2014, **16**, 1.
82. N. Grabowski, H. Hillaireau, J. Vergnaud, N. Tsapis,M. Pallardy, S. Kerdine-Römer and E. Fattal, *International Journal of Pharmaceutics*, 2015, **482**, 75.

Ian Major, Elaine Kenny, Andrew Healy, Luke Geever,
Declan M. Devine and John Lyons

5 Processing of biodegradable polymers

5.1 Introduction

Processing bioresorbable polymers can be achieved *via* conventional polymer processing methods such as extrusion, injection and compressing moulding, solvent spinning or casting. However, special consideration must be given when processing these materials because heat can cause a reduction in molecular weight (MW) due to the hydrolysing of bonds. This chapter outlines the different processes and special considerations needed when processing commonly used biodegradable polymers.

5.2 Hot-melt extrusion

Hot-melt extrusion (HME) is one of the most widely used processing techniques within the plastic and rubber manufacturing industry as well as in the food-processing industry [1, 2]. Over the past few decades, HME has emerged as a powerful processing technology for the production of pharmaceutical solid dosage forms in which an active pharmaceutical ingredient (API) is dispersed into polymer matrices which are generally bioresorbable or at minimum excretable without causing adverse side effects. It has been shown that formulations using HME can provide time-controlled, sustained and targeted drug delivery, and improved bioavailability of poorly soluble drugs [2]

HME can be described as the process of converting a raw material into a product of uniform shape and density by forcing it through a die under controlled conditions [3]. The first extruder is generally accepted to have been built when Joseph Bramah constructed a hand-operated piston press for the manufacture of seamless lead pipes in England in 1797 [4]. Currently, more than half of all plastic products, including plastic bags, sheets, and pipes, are manufactured by this process [5, 6].

In medical applications, extrusion has already been used for years to produce balloon tubing and single- or multi-lumen tubing, to be used for minimally invasive diagnostic and therapeutic procedures [7]. The growing market for medical devices will be partly driven by combination devices that release drugs and manufactured *via* HME processes, including antimicrobial catheters and biodegradable stents releasing anti-clot-forming medication. A multidisciplinary approach is being utilised in the research and development of these products [8].

The HME process involves the transfer of a blend of active substance, polymer and excipients through the heated barrel of an extruder. As implied by the name, in

https://doi.org/10.1515/9783110640571-005

screw extrusion a screw rotates inside the heated barrel. Ram extrusion operates with a positive displacement ram capable of generating high pressures to push the material through the die. Ram extrusion exerts modest and repeatable pressure, as well as a very consistent extrudate diameter. However, a disadvantage of ram extrusion is its limited melting capacity, which causes poor temperature uniformity in the extrudates. These extrudates also have lower homogeneity in comparison with extrudates processed by screw extrusion. For screw extrusion, there is more shear stress and intense mixing. A screw extruder consists of a conveying system, die system, and downstream auxiliary equipment [7].

Screw extrusion can be further differentiated into single-, twin- or multi-screw extrusion. Single-screw extrusion is the most widely used extrusion system. There are three basic functions: conveying, melting and pumping. Single-and twin-screw extruders differ in the mechanism of transport and mixing capabilities [3]. In single-screw extrusion, the transport mechanism is dominated by frictional forces in the solid-conveying zone and by viscous forces in the melt-conveying zone. Single-screw extruders provide for less efficient material dispersion and are generally longer. Conversely, single-screw extruders are mechanically simpler, making them less expensive and giving a higher productivity-to-cost ratio [9].

Twin-screw extruders utilise two screws, usually positioned in parallel. Twin-screw extruders can be a plethora of screw element configurations to control shear and dispersion along the length. Screws can be co- or counter-rotating. Twin-screw extrusion provides for an extensive level of dispersion because mixing occurs at the macroscopic level (material exchange at the screw) as well as at the microscopic level (in areas of high shear) [6]. Twin-screw extruders have lower material residence times to give higher outputs and have a lower tendency of overheating the polymer [9]. In every HME process, polymers are subjected to thermal and shear stress. These are important factors for heat-sensitive products and materials such as biodegradable polymers because they must be forced through a tight screen pack [10]. The temperature can contribute to the depolymerisation of polymer chains, whereas polymer-chain scission may result from the shearing effects of the screw.

A number of strategies have been developed to maintain greater control of the HME process and to provide continual feedback of thermally sensitive materials. Online rheological data can provide information on the macromolecular structure of the molten polymer. Covas and co-workers [11] introduced a concept of online monitoring of polymer rheology during the extrusion process along the extruder. Raman spectroscopy can be used to monitor polymer-melt compositions [12] or to analyse drug content [13]. Furthermore, near-infrared spectroscopy offers real-time information on the physical and chemical properties during continuous processing [14]. An online particle-measurement tool can monitor particle properties such as size and shape [15].

5.3 Injection moulding

Injection moulding is a forming process in which molten polymer is forced at high pressure into a mould, followed by mould cooling and part ejection. The injection moulding of bioresorbable polymers has been used in the manufacture of scaffolds for tissue-engineering applications [16–18]. The injection moulding machine can be identified as: an injection unit; a machine base with hydraulics; a control unit and cabinet; clamping unit with a mould. Moulding machines are categorised by the maximum tonnage of injection pressure available to inject polymer melt into a mould. The injection unit can be horizontal for continuous cycling or vertical for overmoulding applications. Two-shot (or even more) moulding permits the simultaneous moulding of multiple materials. The injection moulding process is not suited to materials with a low melt-flow index, which are usually materials of high MW and high molecular interactions, because successful operation involves the molten polymer to fill the mould cavity prior to solidification. During moulding, the polymer melt is subjected to elevated temperatures, elongational flow, and shear forces.

Injection moulding has been used in various industries to reproducibly create complex parts. The moulding cycle consists of several steps and the most important control parameters are the injection pressure and speed, packing pressure, and mould cooling [19]. Polymer granules enter a single-plasticising screw from a feed hopper in much the same manner as that in single-screw extrusion. Molten polymer collects at the end of the plasticising screw at a set volume known as the 'injection shot'. The screw plunges the injection shot into the mould through a nozzle at a set injection pressure and speed. Upon cooling, the two halves of the mould open and the part is ejected by pins embedded in the mould tool. The mould consists of two halves and opens once the molten polymer has cooled sufficiently to be ejected. The moulding cycle returns to the first step and is repeated. The cycle time of the process is the rate-determining step for production output. Optimum process parameters reduce the cycle time and increase the quality of the product [20].

Injection moulding is an efficient means of producing parts of high tolerance and reproducibility [21] and, therefore, is an excellent process for the production of solid-dosage forms as well as medical and combination devices. Even simple tablets can be injection moulded. Also, moulding provides for improved dust containment and better content uniformity, and can also be considered to be a continuous process [22]. Compound twin-screw extrusion can be used in combination with injection moulding to manufacture drug-delivery systems [23]. The injection moulding process offers extraordinary possibilities to design fine features and thin-gauge walls. True economic advantage can be obtained for production of very large part quantities [24]. Injection moulding offers mass manufacturability of complex shapes; is usable with a diverse range of materials; is accurate even for micro-features [25]. Complex shape forming is highly desirable for tissue-engineering applications, and the growth of moulding of bio-resorbable polymers in this area is strong. Also, the use of the injection mould-ing for autosterilisation of scaffolds is being investigated [26].

5.4 Fused deposition modelling

Three-dimensional (3D) printing, also referred to as 'additive manufacturing', 'rapid prototyping', or 'solid-freeform fabrication', is a process which involves the manufacture of an object or structure by deposition or binding of materials layer-by-layer as shown in [27–29].

Over the last decade, this 30-year old technology, additive manufacturing, has received unprecedented attention and, in particular, the use of 3D printing for biomedical applications. Furthermore, it has been shown to be successful as a method of fabrication for drug delivery. The US Food and Drug Administration recently approved the first 3D-printed medicine, the orally dissolving tablet Spritam® (levetiracetam), marketed by Aprecia Pharmaceuticals, for the treatment of epilepsy in paediatric and geriatric patients who have difficulty in swallowing tablets or who have dysphagia [29–32]. With regard to improvements in the manu-facture of dosage forms, 3D printing allows for the fabrication of multiple drug-containing devices with complex design configurations and release characteristics that would not be possible to fabricate by utilising conventional manufacturing methods.

In recent times, fused deposition modelling (FDM) (Figure 5.1) has been utilised for the fabrication of tablets by pre-fabricating commercial polyvinyl alcohol (PVA) filaments with drugs [33–35].

A group of researchers evaluated the possibility of utilising FDM as a method for fabricating tablets in which they loaded commercial filaments with fluorescein and evaluated how the infill-printing percentage influenced the weight and volume of the printed tablets [36]. The printed tablets were subjected to dissolution testing. The data generated showed that slower release was obtained as the percentage infill in-creased, with the 90% infill taking 20 h for complete release of the active ingredient. The same research team aimed to incorporate the aminosalicylates 5-aminosalicylic acid (5- ASA) and 4-aminosalicylic acid (4-ASA) in commercial PVA filament by en-trapment in an ethanolic solution [33]. The filament strands contained relatively low concentrations of drug, with only 0.06 and 0.25% w/w being achieved for 5-ASA and 4-ASA, respectively. This could be attributed to the low solubility of the API in the solvent (ethanol). The researchers concluded that FDM may be a suitable process for the formulation of tablets containing drugs such as 5-ASA. However, 4-ASA was shown to be less thermally stable at the high printing temperatures required for PVA (210 °C) with a 50% loss of potency in the final printed tablet, which indicated that FDM has limitations with regard to thermally labile molecules. A study by Skowya

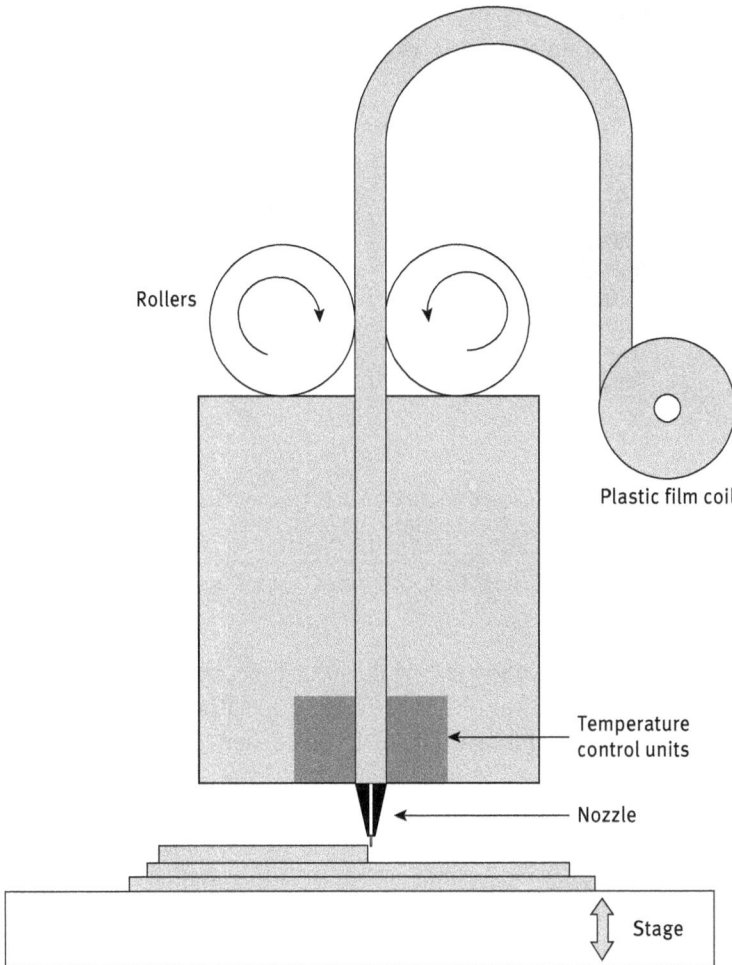

Figure 5.1: Fused deposition modelling 3D printer (schematic). Reproduced with permission from B.C. Gross, J.L. Erkal, S.Y. Lockwood, C. Chen and D.M. Spence, *Analytical Chemistry*, 2014, **86**, 3240. ©2014, Americna Chemical Society [28].

and researchers [34] utilised the same process for encapsulating the drug prednisolone within commercial filaments for the production of tablets. They concluded that it was possible to manufacture a range of different-strength tablets with 88.7–107% of the target potent accuracy, and *in vitro* release data indicated that prednisolone release was extended over 24 h.

More complex designs have incorporated two drugs, caffeine and paracetamol, into commercial PVA filaments by utilisation of HME. This has provided the filaments with inherent characteristics (e.g., a diameter of 1.75 mm) which made it suitable for use in FDM [35]. The filaments were printed using a dual FDM 3D printer into multilayer oral devices, with alternating 1-mm layers of caffeine and paracetamol, and DuoCaplets, a caplet within a caplet, with the inner caplet containing the alternate to that of the outer core (Figure 5.2).

(A)   (B)

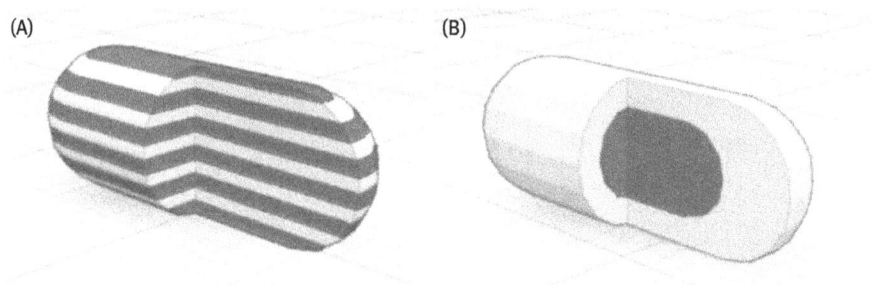

Figure 5.2: Three-dimensional representation of multilayer oral dosage forms: A) sectioned multilayer device (alternating 1-mm layers) and B) sectioned DuoCaplets (caplet within a caplet). Reproduced with permission from A. Goyanes, J. Wang, A. Buanz, R. Martínez-Pacheco,Telford, S. Gaisford and A.W. Basit, *Molecular Pharmaceutics*, 2015, **12**, 4077. ©2015, American Chemical Society [35].

The results generated from dissolution testing illustrated that the difference in the structure of the device has a critical role in the release kinetics. As expected, the drugs were released from the multilayered oral device simultaneously with an increase in drug loading imparting an increase in the release rate of both drugs. The drug release from the DuoCaplets was interesting because the internal caplet released its drug only when the outer layer was completely dissolved. Furthermore, the release profiles were different depending on the drug incorporated in the outer layer. That study highlighted the possibilities of fabricating delayed- or controlled-release devices, which would otherwise not be possible by utilisation of conventional manufacturing methods.

The 3D printing process is not just a method of fabrication of drug-delivery devices or implants. A team of researchers led by Tarcha [37] showed that inkjet-printing technology offers the ability to coat and load stents of different configurations with low doses of drug solutions. They indicated that by utilisation of the aforementioned technology, it is possible to coat stents or other implantable devices with high accuracy and repeatability that may provide a significant improvement in drug-loading efficiency over conventional coating methods [37].

Bioresorbable stents which consisted of a polycaprolactone (PCL) scaffold coated with sirolimus mixed with poly(lactide-*co*-glycolic acid) (PLGA) and polyethylene glycol have been fabricated using a 3D rapid- processing technique similar to FDM [38, 39]. The work which was carried out involved utilisation of a 3D plotting system coupled with an electrospinning apparatus to fabricate the scaffold, and an ultrasonic spray method for the coating application. The studies evaluated the *in vivo* and *in vitro* performance of coated and uncoated stents fabricated *via* the 3D plotting method. Scanning electron microscopy images demonstrated that an uncoated stent exhibited a rough surface finish in comparison with the coated stent, which had a smooth surface, suggesting that the strut surface changed upon coating. *In vitro* drug-release studies indicated that the drug coating remained on the stent and provided additional mechanical support for ≥31 days. *In vivo* studies revealed that stents employing the coating performed more favourably than uncoated counterparts with areas of restenosis. The coating reduced neointimal hyperplasia, inflammation and thrombus formation, underlining how 3D printing could be utilised in the effective production of drug-eluting polymeric stents [39]. For the prevention of biofilms and to stimulate cell development, Gu and researchers [40] inkjet-printed PLGA micropatterns eluting rifampicin and biphasic calcium phosphate nanoparticles onto an orthopaedic implant.

5.5 Special considerations for hot-melt processing bioresorbable polymers

In all hot-melt processing procedures, the polymer is subjected to heat in combination with shear forces and pressure. These tough conditions cause partial degradation of the polymers to affect the MW, distribution, and crystallinity in such a way as to have a negative impact on the polymer properties [5]. Below we outline the thermal properties of the main bioresorbable polymers that are processed *via* melt processing.

5.5.1 Polylactic acid

Polylactic acid (PLA) is an aliphatic biodegradable thermoplastic that can be derived from renewable resources. In recent years, PLA has attracted attention from a research perspective owing to it being renewable, biodegradable and biocompatible as well as exhibiting good mechanical properties [41]. The thermal degradation of PLA-based polymers is initiated by random chain scission or, in some cases, specific chain-end scission, which proceeds *via* the breaking of the carbon–oxygen bond to generate an acid and an unsaturated ester [42]. PLA degradation is driven by four key processes i) hydrolysis; ii) chain depolymerisation; iii) random chain scission by oxidation; and iv) oligomerisation [43, 44]. Pre-drying of PLA granules is an essential first step for the processing of the material. Even then PLA has a narrow processing window which hampers effective melt processing. The glass transition temperature (T_g) and melting temperature (T_m) are 60 and 180 °C, respectively [42]. Rapid polymer degradation is observed under extended isothermal conditions of 240 °C, with substantial mass reduction beginning at 270 °C. The thermal stability of PLA has been shown to have a direct relationship to the ability of the polymer to withstand racemisation [45, 46]. For extended periods at temperatures above 200 °C PLA undergoes oligomerisation and the formation of meso-lactide monomers. The conversion yield is greater the higher the temperature. The addition of chemical compounds has been shown to control the rate of oligomerisation and the type of enantiomer formed [47].

The T_m of PLA is related directly to the optical purity of the polymer [48, 49]. Pure poly(L-lactic acid) (PLLA) is a semicrystalline material with a T_m of ≈180 °C, whereas a decrease in purity of 50% reduces the T_m to <100 °C. Below 50%, the crystallisability is reduced and a single endothermic peak on a differential scanning calorimetry melting curve is no longer readily distinguishable because the polymer is amorphous. The T_g also decreases with increasing D-lactide content [50]. The T_g for pure PLLA is 63 °C and reduces to just above 50 °C for high-content D-lactide grades [51].

The rheological properties of PLA are very much dependent on MW [52, 53]. High-MW PLA melt exhibits pseudoplastic, non-Newtonian flow. At low shear rates, lower-MW PLA melt exhibits Newtonian flow. Increasing D-lactide content decreases the shear viscosity. PLA melt exhibits shear thinning with increasing shear rate [54]. PLA does not exhibit a Newtonian plateau but instead a steep decline in viscosity with increasing frequency at all temperatures observed [42]. Such a phenomenon is related to the decrease in MW wrought by thermal-degradation processes due to the long residence times in the rheometer barrel. Indeed, it is this very characteristic that has limited PLA use in processes that require high melt strength, such as blow-film and blow-moulding extrusion [55–57].

A number of strategies are employed to improve the melt strength of PLA-based materials. One method has been to melt blend the material with other suitable polymers, such as PCL, starch, polybutyrate and polybutylene succinate. A second strategy is to add chain extenders and chain-branching agents to modify the rheological properties of PLA. An early study on PLA clearly demonstrated how chain architecture plays a significant part in PLA flow properties [58]. Chain extenders are reactive short-chain chemical compounds such as multifunctional epoxides [56, 57] and maleic anhydrides (MA) [57].

Injection moulding of PLA is difficult due to the inherent low melt viscosity which restricts the formation of sufficient back pressure for a proper moulding cycle [57]. Injection pressures must also be kept low because shear forces degrade the polymer and, therefore, further compounds the moulding of thin-wall parts. So far, injection moulding has been limited to amorphous PLA grades. Jaszkiewicz and researchers [58] chemically modified commercial grades of PLA resin, one aimed at injection-moulding applications and the other for extrusion processes. Both epoxide and MA compounds were used to chain-extend the resins *via* a reactive extrusion process. Epoxide modification-enhanced chain branching in the injection moulding resin caused a decrease in the flow properties, as observed by a decrease in the spiral flow length. The increased viscosity of the resin necessitated an increase in injection pressure for mould filling.

5.5.2 Polycaprolactone

PCL is an aliphatic polyester and is one of the most important biodegradable polymers used in medicine. PCL-based polymers thermally degrade *via* carbon–oxygen scission mechanisms that are similar to those for PLA-based polymers [42]. However, such polymers have a much broader processing window, showing remarkable thermal stability even at temperatures well in excess of the T_m. The T_g of PCL is −60 °C and the T_m is 60 °C. These low thermal properties provide a high thermal stability in the molten state [59]. During thermogravimetric analysis (TGA), the PCL, while being held at a temperature for 2 h, did not reduce in mass until it was >300 °C. PCL has two thermal-degradation mechanisms [60]. The first begins at 320 °C with the release of water, carbon dioxide, and a carboxylic acid due to the breaking of polyester chains *via* an ester pyrolysis reaction. The second mechanism involves the release of ε-caprolactone monomer, and signifies an 'unzipping' depolymerisation of the PCL. This second mechanism begins at 420 °C.

Due to the low T_g, PCL amorphous architecture displays high molecular motion at physiological temperatures. The mechanics behind the low T_g of PCL is due to the expansion of methylene molecules. This structure makes PCL soft and provides it with a much lower T_g in comparison with polyglycolic acid (PGA). Rheologically, PCL displays typical pseudoplastic behaviour at lower frequencies, with only slight deviation for lower-MW grades [60].

A team of researchers led by Lyons reported extensively on the melt-processing behaviour of PCL and when blended with polyethylene oxide (PEO) [61] in a twin-screw extruder. The researchers used extruder torque as a method of measuring the relative viscosities of polymer melts at set values of processing temperature, feed rate and screw speed. Torque is a measure of motor resistance due to polymer-melt viscosity within the barrel [62]. Lyons and co-authors [61] described several trends from the extrusion compounding processing data of PCL and PCL/PEO blends. Torque and die-head pressure decreased with increasing processing temperatures. Lower values for die-head pressure and torque were noted as a result of thetemperature dependency of polymer melts. Incorporation of PEO resulted in higher melt viscosity during extrusion, giving values for torque and die-head pressure. Steady-state rheometry was used to confirm the lower viscosity of the higher PCL samples. Higher screw speeds resulted in higher values of torque and die-head pressure because higher volumes of polymer melt were processed per unit time. One notable effect of higher screw speeds on PCL was a decrease in the subsequent melt viscosity of the samples.

The authors believe this relates to a decrease in MW to mechanical shearing of the PCL chains, a phenomenon that had been reported previously [63, 64].

5.5.3 Polyglycolic acid

PGA is a semicrystalline aliphatic polyester which has a T_m of 220–225 °C and a T_g of 35–40 °C [65]. This polymer has a very narrow melt-processing window because the onset of thermal degradation begins only 30 °C above the T_m [66]. PGA degrades by random chain scission [67] and chain-end scission [68], with the predominant mechanism being determined by the degradation temperature [69]. Mass reduction is limited at lower temperatures, with the PGA degrading *via* ester-interchange reactions to form larger-olefinic chains. At higher temperatures, there is significant formation of formaldehyde, glycolide and methyl glycolate groups due to chain-end decarboxylation reactions [68, 69]. PLGA is even more widely used than PGA for bioabsorbable applications due to better mechanical performance and ability to tailor biodegradation [70]. Rheologically, PLGA displays Newtonian flow except at high frequencies [71]. PLGA is an amorphous polymer whose T_g increases with an increasing number of lactide groups [72]. PLGA is fairly thermally stable at processing temperatures up to 200 °C [70]. A significant reduction in MW was observed in samples processed at 275 °C. TGA studies have shown that PLGA has a distinct degradation step and that the temperature this steps occurs at increases with increasing lactide content [73]. A PLGA with 50% lactide monomer content degrades at 315 °C, whereas PLA homopolymer starts to degrade at 360 °C. Thermal degradation of PLGA proceeds *via* transesterification reactions that produce cyclic degradation by-products that further degrade the polymer. Increasing the amount of lactide monomer increases thermal stability because the lactide monomer is more resistant to nucleophilic attack.

5.6 Summary

This chapter has discussed production methods that are utilised for biodegradable polymers in the manufacture of biomedical products. Hot-melt extrusion and injection moulding both play key roles in the production of such items. Fused deposition modelling is finding a niche role in the market segment, since it offers advantages over conventional processes, particularly in the production of bespoke or complex parts. However, all three processes involve higher temperatures, shear forces and elevated pressures that can be detrimental to the bulk properties of biodegradable polymers and as such must be given special consideration.

References

1. L. Saerens, C. Vervaet, J.P. Remon and T. De Beer, *Journal of Pharmacy and Pharmacology*, 2014, **66**, 180.
2. M. Stanković, H.W. Frijlink and W.L.J. Hinrichs, *Drug Discovery Today*, 2015, **20**, 812.
3. C. Rauwendaal in *Polymer Extrusion*, 1st Edition, Hanser Publishers, Cincinnati, OH, USA, 1986.
4. P. Lagasse in *The Columbia Encyclopedia*, 6th Edition, Gale Group, New York, NY, USA, 2001.
5. J. Breitenbach, *European Journal of Pharmaceutics and Biopharmaceutics*, 2002, **54**, 107.
6. M.M. Crowley, F. Zhang, M.A. Repka, S. Thumma, S.B. Upadhye, S. Kumar Battu, J.W. McGinity and C. Martin, *Drug Development and Industrial Pharmacy*, 2007, **33**, 909.
7. A.K. Vynckier, L. Dierickx, J. Voorspoels, Y. Gonnissen, J.P. Remon and C. Vervaet, *Journal of Pharmacy and Pharmacology*, 2014, **66**, 167.
8. M. Maniruzzaman, J.S. Boateng, M.J. Snowden and D. Douroumis, *ISRN Pharmaceutics*, 2012, **2012**, 1.
9. N. Follonier, E. Doelker and E.T. Cole, *Journal of Controlled Release*, 1995, **36**, 243.
10. J.L. White and H. Potente in *Screw Extrusion: Science and Technology*, Carl Hanser Verlag GmbH Co. KG, München, Germany, 2002.
11. J.A. Covas, J.M. Maia, A.V. Machado and P. Costa, *Journal of Non-Newtonian Fluid Mechanics*, 2008, **148**, 88.
12. S.E. Barnes, M.G. Sibley, H.G.M. Edwards and P.D. Coates, *Transactions of the Institute of Measurement and Control*, 2007, **29**, 453.
13. V.S. Tumuluri, M.S. Kemper, I.R. Lewis, S. Prodduturi, S. Majumdar, B.A. Avery and M.A. Repka, *International Journal of Pharmaceutics*, 2008, **357**, 77.
14. D. Markl, P.R. Wahl, J.C. Menezes, D.M. Koller, B. Kavsek, K. Francois, E. Roblegg and J.G. Khinast, *AAPS PharmSciTech*, 2013, **14**, 1034.
15. D. Treffer, P.R. Wahl, T.R. Hörmann, D. Markl, S. Schrank, I. Jones, P. Cruise, R.K. Mürb, G. Koscher, E. Roblegg and J.G. Khinast, *International Journal of Pharmaceutics*, 2014, **466**, 181.
16. M. Entezarian, R. Smasal and J.C. Peskar, inventors; Phillips Plastics Corporation, assignee; US7832459, 2010.
17. A. Kramschuster, L-S. Turng, W-J. Li, Y. Peng and J. Peng in *Proceedings of the ASME 1st Global Congress on NanoEngineering for Medicine and Biology* 2010, 7–10th February, Houston, TX, USA, 2010, p.1.
18. J. Vivanco, A. Aiyangar, A. Araneda and H.L. Ploeg, *Journal of the Mechanical Behavior of Biomedical Materials*, 2012, **9**, 137.
19. J. Tiusanen, D. Vlasveld and J. Vuorinen, *Composites Science and Technology*, 2012, **72**, 1741.
20. X-P. Dang, *Simulation Modelling Practice and Theory*, 2014, **41**, 15.
21. L. Wu, D. Jing and J. Ding, *Biomaterials*, 2006, **27**, 185.
22. L. Zema, G. Loreti, A. Melocchi, A. Maroni and A. Gazzaniga, *Journal of Controlled Release*, 2012, **159**, 324.
23. T. Quinten, T. De Beer, C. Vervaet and J.P. Remon, *European Journal of Pharmaceutics and Biopharmaceutics*, 2009, **71**, 145.
24. K-U. Kainer, T. Ebel, O.M. Ferri, W. Limberg, F. Pyczak, F-P. Schimansky and M. Wolff, *Powder Metallurgy*, 2012, **55**, 315.
25. R. Surace, G. Trotta, V. Bellantone and I. Fassi in *New Technologies – Trends, Innovations and Research*, InTech, Rijeka, Croatia, 2012.
26. Z. Cui, B. Nelson, Y. Peng, K. Li, S. Pilla, W-J. Li, L-S. Turng and C. Shen, *Materials Science and Engineering: C*, 2012, **32**, 1674.

27. B.P. Conner, G.P. Manogharan, A.N. Martof, L.M. Rodomsky, C.M. Rodomsky, D.C. Jordan and J.W. Limperos, *Additive Manufacturing*, 2014, **1–4**, 64.
28. B.C. Gross, J.L. Erkal, S.Y. Lockwood, C. Chen and D.M. Spence, *Analytical Chemistry*, 2014, **86**, 3240.
29. L.K. Prasad and H. Smyth, *Drug Development and Industrial Pharmacy*, 2016, **42**, 1019.
30. J. Boetker, J.J. Water, J. Aho, L. Arnfast, A. Bohr and J. Rantanen, *European Journal of Pharmaceutical Sciences*, 2016, **90**, 47.
31. A. Goyanes, H. Chang, D. Sedough, G.B. Hatton, J. Wang, A. Buanz, S. Gaisford and A.W. Basit, *International Journal of Pharmaceutics*, 2015, **496**, 414.
32. G. Jonathan and A. Karim, *International Journal of Pharmaceutics*, 2016, **499**, 376.
33. A. Goyanes, A.B.M. Buanz, G.B. Hatton, S. Gaisford and A.W. Basit, *European Journal of Pharmaceutics and Biopharmaceutics*, 2015, **89**, 157.
34. J. Skowyra, K. Pietrzak and M.A. Alhnan, *European Journal of Pharmaceutical Sciences*, 2015, **68**, 11.
35. A. Goyanes, J. Wang, A. Buanz, R. Martínez-Pacheco, R. Telford, S. Gaisford and A.W. Basit, *Molecular Pharmaceutics*, 2015, **12**, 4077.
36. A. Goyanes, A.B.M. Buanz, A.W. Basit and S. Gaisford, *International Journal of Pharmaceutics*, 2014, **476**, 88.
37. P.J. Tarcha, D. Verlee, H.W. Hui, J. Setesak, B. Antohe, D. Radulescu and D. Wallace, *Annals of Biomedical Engineering*, 2007, **35**, 1791.
38. S.A. Park, S.J. Lee, K.S. Lim, I.H. Bae, J.H. Lee, W.D. Kim, M.H. Jeong and J-K. Park, *Materials Letters*, 2015,**141**, 355.
39. S.A. Park, H.J.H.K. Kim, S.H. Lee, J.H. Lee, H.J.H.K. Kim, T.R. Yoon and W. Kim, *Polymer Engineering & Science*, 2011, **51**, 1883.
40. Y. Gu, X. Chen, J.H. Lee, D.A. Monteiro, H. Wang and W.Y. Lee, *Acta Biomaterialia*, 2012, **8**, 424.
41. Y. Chen, L.M. Geever, J.A. Killion, J.G. Lyons, C.L. Higginbotham and D.M. Devine, *Polymer Composites*, 2015, DOI:10.1002/pc.23794.
42. A.L. Arraiza, J.R. Sarasua, J. Verdu and X. Colin, *International Polymer Processing*, 2007, **22**, 389.
43. F-D. Kopinke, M. Remmler, K. Mackenzie, M. Möder and O. Wachsen, *Polymer Degradation and Stability*, 1996, **53**, 329.
44. A. Södergård and M. Stolt, *Progress in Polymer Science (Oxford)*, 2002, **27**, 1123.
45. T. Tsukegi, T. Motoyama, Y. Shirai, H. Nishida and T. Endo, *Polymer Degradation and Stability*, 2007, **92**, 552.
46. L.T. Lim, R. Auras and M. Rubino, *Progress in Polymer Science*, 2008, **33**, 820.
47. Y. Fan, H. Nishida, Y. Shirai and T. Endo, *Green Chemistry*, 2003, **5**, 575.
48. D.M. Bigg in *Proceedings of ANTEC ' 96:Plastics-Racinginto the Future*, 5–10th May, Indianapolis, IN, USA, 1996, p.2028.
49. J-R. Sarasua, R.E. Prud'homme, M. Wisniewski, A. Le Borgne and N. Spassky, *Macromolecules*, 1998, **31**, 3895.
50. H. Urayama, T. Kanamori, K. Fukushima and Y. Kimura.
51. J.R. Dorgan, J. Janzen, M.P. Clayton, S.B. Hait and D.M. Knauss, *Journal of Rheology*, 2005, **49**, 607.
52. J.J. Cooper-White and M.E. Mackay, *Journal of Polymer Science, Part B: Polymer Physics*, 1999, **37**, 1803.
53. J.R. Dorgan, H. Lehermeier and M. Mang, *Journal of Polymers and the Environment*, 2000, **8**, 1.
54. Q. Fang and M.A. Hanna, *Industrial Crops and Products*, 1999, **10**, 47.
55. R. Al-Itry, K. Lamnawar and A. Maazouz, *Polymer Degradation and Stability*, 2012, **97**, 1898.
56. R. Al-Itry, K. Lamnawar and A. Maazouz, *Polymers*, 2015, **7**, 939.
57. A. Jaszkiewicz, A.K. Bledzki and A. Meljon, *Polymer Degradation and Stability*, 2014, **101**, 65.

58. A. Jaszkiewicz, A.K. Bledzki, A. Duda, A. Galeski and P. Franciszczak, *Macromolecular Materials and Engineering*, 2014, **299**, 307.
59. P. Yilgor, R.A. Sousa, R.L. Reis, N. Hasirci and V. Hasirci in *Macromolecular Symposia*, 2008, **269**, 92.
60. O. Persenaire, M. Alexandre, P. Degée and P. Dubois,*Biomacromolecules*, 2001, **2**, 288.
61. J.G. Lyons, P. Blackie and C.L. Higginbotham, *International Journal of Pharmaceutics*, 2008, **351**, 201.
62. G. Verreck, A. Decorte, H. Li, D. Tomasko, A. Arien, J. Peeters, P. Rombaut, G. van den Mooter and M.E. Brewster, *Journal of Supercritical Fluids*, 2006, **38**, 383.
63. D. Homminga, B. Goderis, S. Hoffman, H. Reynaers and G. Groeninckx, *Polymer*, 2005, **46**, 9941.
64. J.G. Lyons, M. Hallinan, J.E. Kennedy, D.M. Devine, L.M. Geever, P. Blackie and C.L. Higginbotham, *International Journal of Pharmaceutics*, 2007, **329**, 62.
65. S.M. Davachi, B. Kaffashi, J.M. Roushandeh and B. Torabinejad, *Materials Science and Engineering: C*, 2012, **32**, 98.
66. O. Dechy-Cabaret, B. Martin-Vaca and D. Bourissou,*Chemical Reviews*, 2004, **104**, 6147.
67. K. Chujo, H. Kobayashi, J. Suzuki and S. Tokuhara, *Die Makromolekulare Chemie*, 1967, **100**, 267.
68. I.C. McNeill and H.A. Leiper, *Polymer Degradation and Stability*, 1985, **12**, 373.
69. G. Sivalingam and G. Madras, *Polymer Degradation and Stability*, 2004, **84**, 393.
70. M.S. Widmer, P.K. Gupta, L. Lu, R.K. Meszlenyi,G.R.D. Evans, K. Brandt, T. Savel, A. Gurlek, C.W. Patrick,A.G. Mikos and W.P.J. Charles, *Biomaterials*, 1998, **19**, 1945.
71. J. Aho, J.P. Boetker, S. Baldursdottir and J. Rantanen, *International Journal of Pharmaceutics*, 2015, **494**, 623.
72. M. Penco, L. Sartore, F. Bignotti, S. D'Antone and L. Di Landro, *European Polymer Journal*, 2000, **36**, 901.
73. S. D'Antone, F. Bignotti, L. Sartore, A. D'Amore, G. Spagnoli and M. Penco, *Polymer Degradation and Stability*, 2001, **74**, 119.

Emily Crowley, Maurice Dalton and Gavin Burke

6 Cytotoxicity and biocompatibility of bioresorbable polymers

6.1 Introduction

All materials implanted into living tissue initiate various tissue responses extending from injury to recruitment of immune cells. This results in inflammation and responses that allow wound healing at the site of implantation [1].

Biocompatibility is greatly influenced by the biological environment and also the various exclusive interactions that occur between different materials and the tissues that they come into contact with [2].

A major concern of biomaterial research is the extent of interaction after implantation between the biomaterial and the host. Therefore, biomaterials must undergo extensive assessment of their biocompatibility to ensure that the benefits of implantation outweigh negative outcomes. These studies include multifaceted *in vitro* and *in vivo* experiments to determine the possible effects of implantation. *In vitro* analysis, *via* cell culture tests, allow the prediction of potential tissue responses at a specific site. Biocompatibility testing of biomaterials is essential because the functionality and robustness of an implantable device may be affected as a result of the host's reaction to the foreign substance [3].

Materials used for biodegradable polymeric implants must undergo safety assessment before being utilised [4] because these compounds are in direct contact with cells and tissues within the human body [5, 6]. Compounds implanted within the body must be non-toxic and non-injurious to prevent rejection after implantation. A biocompatible implant is one whose advantages outweigh the reaction caused upon implantation and has the ability to break down into smaller, safer constituents that will ultimately be eradicated from the body with no trace. Many factors affect the biocompatibility of biomaterials. Parameters such as implant size, design, morphology and degradation play an important part in the biocompatibility of biomaterials [7]. The outcome after implantation of the biomaterial is dictated by a combination of factors, with the implant site and microenvironment having primary roles. A multifaceted procedure involving *in vivo* and *in vitro* tests, and biocompatibility testing of polymers aims to evaluate the potential adverse effects of the biomaterials used in drug delivery. Evaluation of the biocompatibility and safety of bioresorbable polymers incorporates many parameters to highlight the effects of these materials on implantation.

Methods such as histopathological examination after material implantation and *in vitro* analysis of the effects to cells are done [5, 8]. Genotoxic assessment of biomaterials must also be carried out to assess the biocompatibility of the intended implant.

https://doi.org/10.1515/9783110640571-006

Many factors affect the formation of the adverse reaction after polymer implantation, and include biomaterial composition and degradation rate. Implant size, design and the structure of the implant surface affect the rate of degradation of the biomaterial [7]. Blood flow, implantation site and patient information also affect this rate. A faster resorption rate can be observed in cases where the biomaterial is under stress and causes implant breakage, resulting in an increased surface area. Toxicology, biocompatibility and biodegradability studies are essential for all polymers utilised for clinical because one tissue may react differently to another [8].

Owing to reduced/absence of toxicity, the composition of a synthetic polymer can be controlled during production. Artificial polymers utilised in biomedical applications are degraded into metabolites *via* the Krebs cycle, enzymatic processes and urinary excretion, resulting in elimination from the site of implantation [9–11]. In addition, phagocytosis and/or solubilisation in natural bodily fluids also contribute to the breakdown of small-molecular weight (MW) polymer fragments at the site of implantation. With advantages such as controlled degradation rate and mechanical properties of the biomaterial for biomedical applications and mass generation capabilities, synthetic polymers are favoured over natural biodegradable polymers. Eliminating the need for implant removal, biopolymers such as polylactic acid (PLA), polycaprolactone (PCL) and polyglycolic acid (PGA) are commonly used because they breakdown in the body over time, during which time new tissue is generated [12, 13].

The use of *in vivo* models to assess the biocompatibility of biomaterials intended for use in wound healing is essential because it identifies reactions of the human wound-healing process. *In vitro* models, such as various cell lines representing different sections of the human body, are imperative to determine the cytotoxicity of biomaterials intended for use in guided tissue regeneration (GTR) but it is essential to investigate the biocompatibility of biomaterials in *in vivo* models also because it provides an indication of the human response to the biomaterial [14].

6.2 Foreign-body reaction

The implantation of biomaterials always results in a foreign-body reaction (FBR). The generation of an inflammatory reaction aims to protect tissue at the site of implantation from damage, insulate and remove the foreign material, and also to initiate the intended repair mechanism [7, 15]. Also known as a FBR, the inflammatory process generated post-biomaterial implantation can be divided into two phases: acute and chronic (Figure 6.1). Short-term inflammation is advantageous for the healing process because many important growth factors are produced and it aids in the elimination of unwanted microorganisms from a particular site [16]. Prolonged or chronic inflammation, on the other hand, may cause the inflamed site to enter a chronic state where the intended wound healing is inhibited [17].

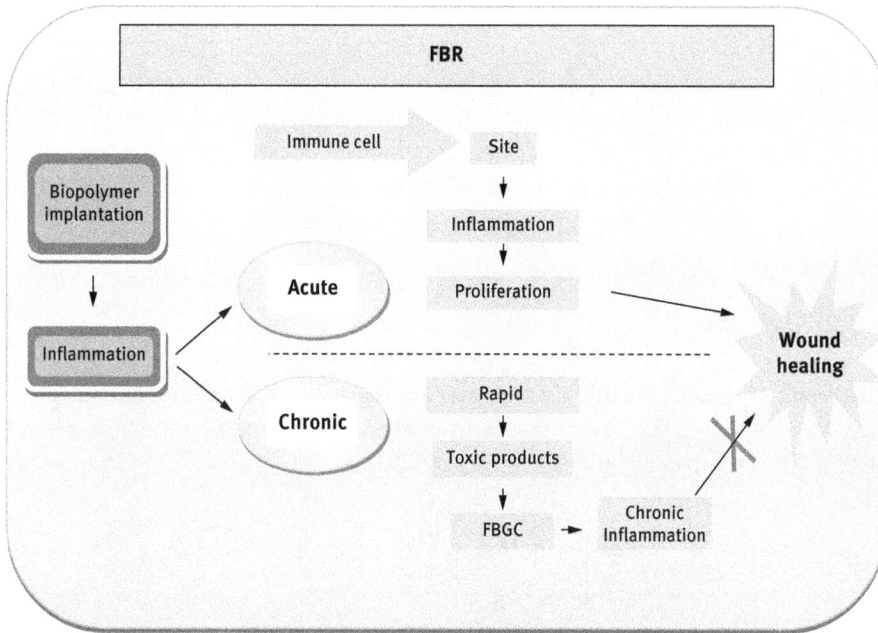

Figure 6.1: A FBR post-biomaterial implantation resulting in acute or chronic inflammation whereby acute inflammation aids in the wound-healing mechanism. On the other hand, chronic inflammation can inhibit the intended wound healing.

The degradation processes of the polymeric biomaterials utilised may affect the tissues surrounding the implantation site. Various toxic reactions, such as inflammation, may occur after the first phase of degradation when the chemical products of the biomaterial degradation process are released from the implant. Studies have demonstrated that slow degradation of biomaterials releasing relatively non-toxic degradation products do not cause adverse reactions in the body. On the other hand, if biomaterial degradation is rapid and resulting in the generation of toxic degradation products, then this may lead to chronic inflammation in the body and generate systemic reactions [12].

In the acute phase of the FBR a vast amount of cells are involved. These cells cooperate, resulting in the generation of a chronic-phase reaction. In a FBR, macrophages 'manage' a complicated cellular reaction. Macrophages are the principal cells in inflammation at the site of implantation. In their aim to phagocytose the foreign material, macrophages become foreign body giant cells (FBGC) that capture the implanted material to protect the body, resulting in biomaterial degradation over time. This reaction to biomaterials can lead to ineffective tissue engineering. The immune reaction generated after implantation of a foreign material determines acceptance of the biomaterial by the patient [8].

6.3 Natural biomaterials

One of two types of biodegradable polymers is natural-based polysaccharides. This group of polymers includes materials such as alginate, chitosan (CS) and derivatives of hyaluronic acid. These polymers generally enhance the interaction between the implanted biomaterial and the host tissue because a specific area of the polymer aids in cell support and generation during development [18].

6.3.1 Alginate

As a result of its biocompatibility and low toxicity, alginate, a natural- anionic polymer, has been used in a vast amount of biomedical applications, such as wound healing and bioactive-agent delivery (Figure 6.2) [18].

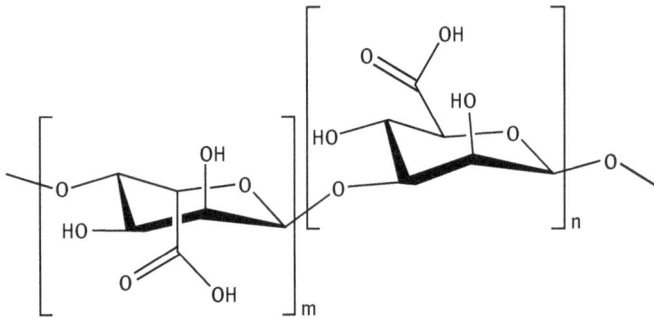

Figure 6.2: Chemical structure of alginate.

Alginate is a water-soluble anionic polysaccharide extracted from the cell walls of brown sea weed (*Laminaria hyperborea*, *Ascophyllum nodosum* and *Macrocystis pyrifera*) and two bacterial genera (*Pseudomonas* and *Azotobacter*) [19, 20]. This copolymer possesses a variety of mixed salt cations such as Mg^{2+}, Sr^{2+}, Ba^{2+} and Na^+ [20]. Its composition includes covalent linked blocks of (1–4)-α-L-guluronic (G) and β-D-mannuronic (M) acid residues. These blocks are composed of consecutive G and M residues (GMGMGM). This composition has a key role in its unique properties, such as stability, size, biocompatibility and permeability. As a result, alginate has been used as a biomaterial for several applications, including tissue- engineering scaffolds, drug-delivery systems and wound dressing [21]. For example, Silva and co-workers investigated the use of CS and alginate layer-by-layer (LbL) deposition to control the release rate of diclofenac *via* silicone-based hydrogels. They demonstrated that the coating could control diclofenac release for several days due to its hydrophilicity and biocompatibility [22]. Raguvaran and his research team utilised sodium alginate/gum acacia and zinc oxide nanoparticles (NP) for wound regeneration and bacterial inhibition. They demonstrated that low concentrations of zinc oxide NP treated on cultures of *Pseudomonas aerigunosa* and *Bacillus cereus* and its biocompatibility on peripheral blood mononuclear/fibroblast cells improved the healing effect [23].

However, alginate biocompatibility has been investigated extensively and its biocompatibility is controversial. Induction of a FBR and fibrosis has been reported, whereas other research reports have shown little or no immune response around alginate implants. In addition, the immunogenicity of the uronic acids in alginates is controversial [24]. There have also been reports that alginates with high M content evoke an inflammatory response so that monocytes are stimulated to produce cytokines such as interleukin (IL)-1, IL-6 and tumour necrosis factor (TNF). It has been suggested that this mechanism may be driven by cluster of differentiation (CD)14 binding [25]. Tam and co-workers compared the biocompatibility of two commercially available alginates; one containing 71% guluronate (HiG), and the other containing 44% (IntG). They suggested that alginate biocompatibility is influenced by mannuronate/guluronate content and intrinsic viscosity [26].

6.3.2 Chitosan

Similar to alginate, CS has many desirable properties: it is non-toxic, biocompatible and biodegradable and aids the proliferation of cells, which makes it an attractive polymer for tissue engineering [2]. CS is a linear polysaccharide comprising random β-(1–4)-linked D-glucosamine (deacetylated unit) and N-acetyl-D-glucosamine (acetylated unit) (Figure 6.3). CS has been utilised in various biomedical fields, including sutures, wound dressing, and drug delivery, owing to its excellent properties: it is non-toxic, biocompatible and can mimic the natural environment [27]. *In vitro* studies have reported favourable derivatives within CS that have the potential to induce growth of many cell types, such as endothelial cells and

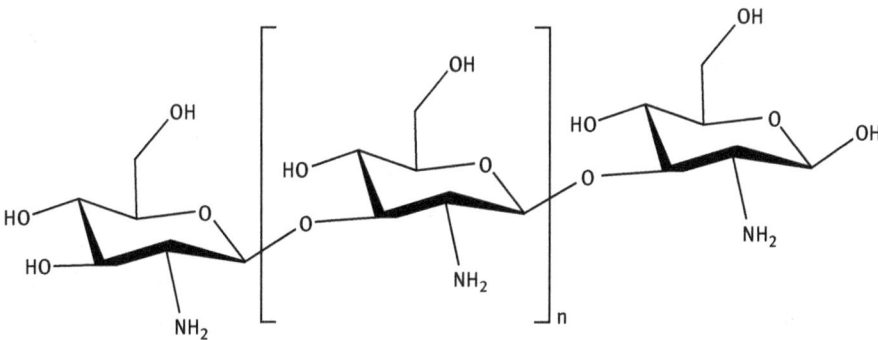

Figure 6.3: Chemical structure of CS.

chondrocytes. The properties, biodegradability and biological role of CS are dependent on the proportions of β-(1–4)-linked D-glucosamine and N-acetyl-D-glucosamine residues [28, 29]. Other intrinsic properties include antibacterial and antifungal activity, as well as mucoadhesive and haemostatic features. The mucoadhesion is due to a protonated amino group along D-glucosamine. The mechanism of action is due to negatively-charged glycoprotein residues within mucus. In acidic media, the CS amino groups are positively charged and can interact with mucin. The haemostatic activity is directly related to the presence of positive charges on the CS backbone. Red blood cell membranes are negatively charged, thereby allowing interaction between CS and red blood cells. CS with protonated amino groups becomes a polycation that allows the polysaccharide to form ionic complexes that can be natural or synthetic, such as lipids, proteins, deoxyribonucleic acid and PGA. The ability to act as a polyelectrolyte allows CS to be used for multilayer films *via* LbL deposition [28]. Venkatesan and co-workers developed a porous CS-alginate and silver NP nanocomposite with antibacterial and anticancer properties. They concluded that the composite was effective in controlling pathogens and had anticancer properties against breast cancer cells [29]. Another important factor for controlling the rate of CS degradation is the degree of acetylation and the alteration of side groups, i.e., extensive hydrogen bonding is prevented by attaching bulky side groups [30]. In addition, the increase in deacetylation provides more positively-charged amino acids, which is a mechanism exploited in drug-delivery systems. For example, highly-deacetylated CS is used for the interaction of negatively-charged residues of mucus, namely sialic acid. This mucoadhesion is responsible for prolonged retention at the site of action and absorption [31]. Furthermore, CS has been employed to aid wound regeneration. As a biopolymer it is known in the wound management field for its haemostatic properties. It also possesses biological activities to stimulate cell proliferation and affects macrophages.

6.3.3 Hyaluronic acid

Occurring naturally, hyaluronic acid is present in all vertebrates and has 'unique viscoelastic properties'. Due to its properties, it is an ideal contender in the field of tissue repair and regeneration because it has good compatibility and degradability within the human body [2]. Hyaluronic acid is distributed widely in the extracellular material (ECM) of connecting tissue [32]. This non-sulfated glycosaminoglycan comprises recurring disaccharides moieties of N-acetyl-D-glucosamine and D-glucuronic acid (Figure 6.4). It plays an important part in the regulation of cell adhesion, tissue morphogenesis and modulation of inflammation. Hyaluronic acid has been a popular biomaterial and has been used in various applications from tissue-culture scaffolds to cosmetic materials. This is due to its unique characteristics such as biocompatibility, biodegradability, non-toxicity and hydrophilicity [33]. Due to its highly flexible nature,

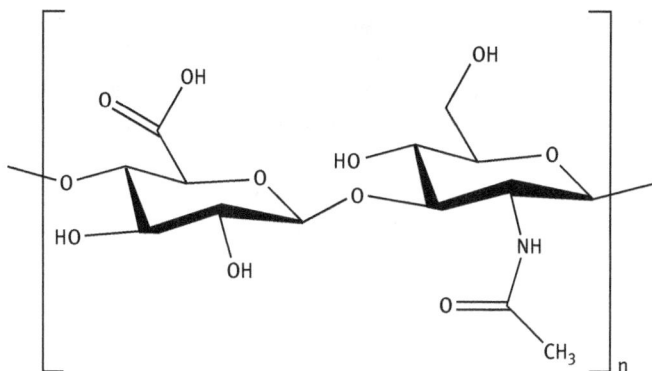

Figure 6.4: Chemical structure of hyaluronic acid.

hyaluronic acid can assume different shapes and configurations to improve its mechanical characteristics. For example, it can be used to improve the properties of other polymers. The disadvantages of using collagen for biomedical applications are its low-biomechanical stiffness and rapid degradation. However, incorporation of hyaluronic acid not only improves the viscoelastic properties of the polymer, it is resistant to deformation and has a high affinity for osteoblasts and chondrocytes, enabling it to be used for scaffold preparation. Sionkowska and his research team utilised hyaluronic acid to improve a three-dimensional composite comprising CS and collagen. They noticed mechanical and thermal stability while incorporating hyaluronic acid [34]. The mechanism of hyaluronic acid *in vivo* is recognised by cell-surface receptors such as CD44, the receptor for hyaluronan-mediated motility. The mechanism of this biocompatibility is based on hyaluronic acid binding to these receptors and resulting in activation of intracellular signalling pathways or cellular uptake [35]. In addition, Wang and co-workers developed novel poly(hyaluronic acid)-doped poly(3,4-ethylenedioxythiophene)/HA/poly(L-lactic acid) (PLLA) films as conductive matrices for nerve-tissue engineering. They demonstrated that, by combining PLLA and hyaluronic acid, there was an increase in neurite growth and expression of messenger ribonucleic acid *via* PC-12 cell lines, thereby indicating a promising strategy for nerve repair [36]. Hyaluronic acid also plays a pivotal part in promoting cell motility and differentiation in wound healing. Blocking of the interaction of hyaluronic acid with CD44 receptors in pre-chondrogenic micromass cultures from embryonic limb bud mesoderms has been shown to inhibit chondrogenesis. This maintenance of differentiated chondrocytes is dependent on interaction between hyaluronic acid receptors [37]. Hyaluronic acid possesses a large spectrum of MW, and exhibits different properties during remodelling of endogenous tissue. Essentially, high-MW hyaluronic acid is a natural barrier for migration and cellular proliferation of endothelial cells. Low-MW hyaluronic acid has been shown to up-regulate the proliferation, migration and sprout extension of endothelial cells. For example, Khanmohammadi and co-workers showed that enzyme immobilisation of low-MW hyaluronic acid within a gelatin hydrogel promoted the mobility of endothelial cells for vascular-dense tissue [38]. In addition, Xu and his research team developed a novel delivery system for the treatment of malignant glioma comprising curcuminoid (Cur)-loaded polysaccharide nanoformulations that could cross the blood–brain barrier (BBB). Penetration and brain-targeting properties were based on hyaluronic acid and chitosan hydrochloride lactoferrin-Cur-polysaccharide NP. They concluded that the delivery system was safe when using brain capillary endothelial cells and C6 cells. However, it also travelled across the BBB to exhibit strong toxic effects on glioma cells [39].

6.4 Synthetic biomaterials

6.4.1 Polylactic acid

PLA is a 'thermoplastic aliphatic polyester' [12] derived entirely from renewable reserves. PLA is one of the most profuse US Food and Drug Administration (FDA)-approved bioresorbable polymers [40]. α-Hydroxy acid-derived polyesters such as PLLA, poly(D-lactic acid) and poly(D,L-lactic acid) (PDLLA) are the most commonly used PLA materials for temporary devices used in biological systems [41]. PLA is readily biodegradable and has tremendous commercial aptitude for bioplastics. Some properties of PLA restrict its use in a variety of applications and, therefore, attempts have been made through co-polymerisation or physical blending with other polymers to enhance its toughness [40]. The arrangement of units within the polymer determines its stability, and polymer degradation rate can be reduced by adapting the principal relevant properties of the polymer, such as MW and crystallinity. Studies have demonstrated the advantages of using PLA and its copolymers for clinical use. It is of utmost importance to ensure that the original properties of the biomaterial being utilised for implantation do not become altered and result in changes in the degradation rate to cause a FBR after implantation [8].

PLA can be formed *via* the fermentation of glucose in corn or sugarcanes. With its attractive properties, such as the capability to degrade in the body and advantageous mechanical properties, it is one of the most commonly studied biodegradable synthetic polymers. The mechanical properties of PLA can be adjusted by altering the proportion of D-lactic acid to L-lactic acid [12, 42].

Simple hydrolysis allows the polymer to degrade followed by elimination of the products through metabolic pathways within the body [41]. The principal degradation process that lactic acid-based polymers undergo is ester hydrolysis, which results in the formation of smaller, water-soluble constituents that can be broken down further *via* metabolism followed by removal from the body.

PLA degradation results in the formation of lactic and glycolic acids, thereby increasing the potential risk of local acidic toxicity at the site of implantation. Generally, bodily fluids have the ability to neutralise the acidity of the degradation products formed. Lactic acid products formed from PLA degradation become incorporated into the tricarboxylic acid cycle and are then excreted from the body. All living cells utilise this cycle within the body to generate the energy they require to grow and divide. It is also known as the 'Krebs cycle' or the 'citric acid cycle' and occurs within mitochondria in cells [43]. Rapid degradation of PLA and slow elimination of acidic products results in accumulation of acidic products within the body, which may induce inflammatory complications. The polymer degradation rate dictates the accumulation of by-products within the body. The degradation rates of biomaterials are influenced by various factors, including the size and design of the biomaterial as well as factors associated with the host, such as the site of implantation and patient criteria [8]. In the course of ester hydrolysis, the terminal ends of carboxylic chains catalyse the breakdown of ester bonds in PLA, resulting in its degradation to release lactic acid [12].

As a filler device for skin applications, PLA is an approved polymer according to the FDA. With a diverse extent of uses in medicine, including drug-delivery systems, resorbable sutures and implants, PLA has desirable and undesirable properties affecting its use. Although it is advantageously biocompatible, PLA usage may be restricted due to its brittle nature and slow crystallisation. Lactic acid polymers characterise a major group of biodegradable polymers in clinical use. In Europe and the USA, PLLA is utilised for soft-tissue growth [8]. Synthetic polymers such as PLA are desirable scaffolds in tissue engineering because they aid the tissue-remodelling process while maintaining their mechanical intensity [42, 44].

Studies carried out in humans to determine the effect of PLA implantation on the immune system have demonstrated that the FBR is initiated with a recruitment of FBGC and macrophages to the site of implantation followed by collagen production. Findings of PLA reactions in rat tissue have shown that, shortly after PLA implantation, a dense cell layer of macrophages is formed followed by the generation of a connective-tissue capsule that surrounds the implant. These studies have also shown that PLLA resorption does not induce an inflammatory reaction and that, during this phase of resorption, cellular activity is restored. Using Wistar Albino Glaxo rats as an *in vivo* model, it was revealed that subcutaneous implantation of PLA resulted in an inflammatory response that decreased after 6 months followed by an evident increase after 12 months due to PLA fragmentation [45]. Damage to surrounding tissue at the site of implantation was noted as a side effect of PLA degradation due to the formation of sharp edges of the implant, and concluded in an inflammatory response. *In vitro* studies using various forms of PLA have illustrated the effect of the by-products of PLA degradation on pH and cell viability. A reduction in cell viability has been observed in cartilage cells exposed to PLA-degradation products. When comparing both by-products, it has been shown that glycolic acid has a more cytotoxic effect than lactic acid [8]. Assessment of PLA implants has shown no genotoxic potential. Adverse reactions after implantation are influenced by the composition of the implanted biomaterial and its degradation rate *in vivo*. With a slower degradation rate than that observed for other polymer implants, PLA reactions occur as late as 2–3 years after implantation [8]. Liu and co-workers studied the effect of nano-hydroxyapatite (nHA) filler on the degradation behaviour of PLA at body temperature in a neutral buffer. Their results illustrated that the weight of PLA did not alter after 50 days of immersion. In the same study with PLA/ 18% nHA composites, there was a proportional degradation of the composite with immersion duration, thereby highlighting the influence of the filler on the degradation behaviour of PLA by increasing the hydrophilicity of the polymer matrix.

Cytotoxicity and bioactivity analysis carried out by Liu and co- workers on PLA composites incorporating nHA filler demonstrated the need for nHA incorporation for the adhesion and attachment of bone cells. These results highlight the inability of bone cells to attach to bio-inert PLA compared with PLA/nHA composites. The addition of nHA filler results in suitable bioactivity and biocompatibility to PLA. Liu and co-workers also demonstrated the cytotoxic nature of PLA/nHA composites exposed to this composite using the MTT assay (which measures the mitochondrial activity within cells). Results emphasised the enhancement of cell growth compared with that of PLA alone although, after several days, cellular viability was reduced significantly as a result of the cytotoxic influence of the silver NP introduced. Here, osteoblast proliferation and growth was diminished. Therefore, low-loading silver levels maintained the advantageous biocompatibility of PLA [46].

In studies completed by Rajmohan and co-workers, the initial outcomes of PLA biocompatibility in guinea pigs as an *in vivo* model demonstrated no adverse reactions upon subcutaneous implantation of the PLA membrane. Biocompatibility evaluation of porous PLA membranes utilised for wound healing must provide information to ensure they are safe for human use. As well as being clinically effective, PLA membranes in that study were put through specific tests outlined in International Organization for Standardization (ISO) 10993 for the assessment of medical devices to demonstrate that these membranes did not inhibit cell growth and proliferation and were non-toxic *in vitro*. Histological studies and visual inspection at the site of implantation in the *in vivo* model showed that no major adverse effect was induced upon implantation of the guinea pig model with the PLA membranes. When comparing the protein and ECM profile between the PLA-implanted guinea pig and control, no effects on subcellular levels post-implantation were observed. PLA membranes have also been shown to be non-toxic towards B16 melanoma cell lines. These melanoma skin cells derived from a mouse model readily adhered and proliferated during these PLA biocompatibility tests [47].

Analysis carried out by Gopimohan and co-workers on the cytotoxicity of drug-loaded PLA for GTR showed no evidence of severe inflammation induced by subcutaneous implantation of the membrane in the *in vivo* model utilised (guinea pig) [48].

Mohiti-Asli and co-workers demonstrated that ibuprofen-loaded PLA nanofibrous scaffolds enhanced the viability of exposed human embryonic kidney (HEK) cells in their *in vitro* cytotoxicity analyses. After enhancing HEK proliferation over a 14-day period of exposure, ibuprofen addition to the PLA scaffolds showed greater viability than HEK cells exposed to pure PLA alone. Cytotoxic analysis was also carried out on another *in vitro* model: primary human dermal fibroblasts (HDF). Generally, cells exposed to ibuprofen-loaded PLA or control remained viable. Also, the specific weights of ibuprofen added enhanced cell proliferation in comparison with the others. In the same study, *in vivo* analysis was completed on the same scaffolds to determine the response of keratinocytes and fibroblasts. Over time after scaffold implantation, degradation began to result in scaffold breakage *in vivo*. Histological examination of the scaffold 14 days post-implantation in nude mice demonstrated the ability of skin cells to migrate through the scaffold. Surfaces of the PLA/ibuprofen scaffolds were surface-treated using atmospheric plasma to increase the hydrophilicity of the composite scaffolds, and resulted in greater cell adhesion which led to improved cell viability and growth of the cells in response [49]. Shah and co-workers demonstrated that endothelial cells introduced to gas plasma-treated PLA films showed greater attachment, viability and proliferation [50].

Biodegradation examination of the ibuprofen-loaded PLA scaffolds showed reduced degradation after the 2-week experimentation period. This finding emphasised the benefit of utilising this scaffold in acute and chronic wound healing because it has the ability to provide the mechanical support needed to initiate the wound-healing process and also for long-term drug release at the site of implantation. Analysis carried out to observe the effect of PLA membranes on human keratinocytes showed that these cells were capable of proliferating on the scaffolds examined. *In vivo* analysis was also completed on the same membranes showing cell migration of HDF and HEK cells throughout the implanted PLA scaffolds. The PLA scaffolds became degraded after 2 weeks of implantation, which highlighted the advantage of using these scaffolds for wounds that demand initial mechanical support [49].

After exposing osteoblasts (cells that construct bone) to PLLA, extracts showed a reduction in cell growth and proliferation with a decrease of pH over time [8, 51]. To enhance bone growth and to aid healing between bone and tendon, combinations of PLLA with slow-dissolving oxides or with bioactive nHA are utilised to overcome the acidic environment formed by PLLA alone [52, 53]. Hydroxyapatite (HA) aids cyto-compatibility because it is a building block of bone and has a buffer effect *in vitro* to adjust the acidic *milieu* caused by PLLA degradation products [54].

A FBR, such as an inflammatory reaction, can be observed after breakdown of PLLA bioresorbable screws. Symptom respite can be achieved through removal of cysts and screw fragments at the site of implantation [55]. PLLA implantations have resulted in FBR such as implant-site swelling ≈3 years after surgery. The tissue reaction to PLLA implants incorporates the formation of a fibrous capsule composed of mostly fibrocytes, which surround the foreign material. Approximately 5 years after implantation of PLLA biomaterial, microscopic evaluation shows a thin fibrous capsule with blood and nerve supplies as well as fat deposition on the implant itself. Cells engulfing the foreign material become structurally damaged. With increasing time of implantation, there is an increase in macrophage and FBGC production and a reduction in the number of fibrocytes. PLLA does not have the capacity to cause detrimental cell injury resulting in cell death. In separate studies, it has been shown that biomaterials such as PLLA utilised for the treatment of lesions of the labrum or rotator cuff can lead to adverse reactions. Inflammatory reactions have been observed from PLLA degradation products, with mechanical damage to bone cells being the principal effect after implantation [56].

In most cases, the degradation products of the polymer generate the inflammatory response. Carcinogenesis results from long-term minor inflammation. This causes a flow of events that result in tumour formation. Implantation of PLLA biomaterial in rodents has been shown to result in sarcoma formation at the site of implantation with no systemic effects of the tumour in a rodent model [8]. PLLA microspheres have been shown to support chondrocyte growth if grown in a bioreactor with polymer surface modification of minor repetitive amino acid sequences. An amplified growth of cells was observed by utilisation of a bioreactor and a positive capacity for cell adhesion and proliferation was observed in one study [41].

The properties and degradation rates of various PLA forms vary with different biomaterial structure and synthetic configuration. Li and co-workers established through *in vitro* and *in vivo* analysis in rabbits, rats and mice that exposure to PDLLA instigated no skin irritation and that PDLLA was non-cytotoxic because there was no significant reduction in Vero cell viability or morphology when compared with cells cultured in optimum conditions. After those results, Li and co-workers carried out a micronucleus test to examine the potential genotoxicity of PDLLA films. Results indicated that PDLLA films did not have the capacity to alter formation of micronucleated polychromatic erythrocytes, and proposed that PDLLA films are non-genotoxic in mice models. That study indicated that PDLLA films have reduced toxicity within the human body. Xiang and co-workers carried out studies to show the variation of cell adhesion and proliferation of MC3T3-E1 cells on a modified-PLA biomaterial compared with a PDLLA surface. Results demonstrated that maleic anhydride modified-PLA biomaterials aided cell adhesion and spread to a greater extent than PDLLA because addition of carboxylic acid side chains to the biomaterial were correlated with effective MC3T3-E1 cell attachment and spread. *In vivo* implantation of the PDLLA film resulted in no alteration of haemoglobin or platelet counts, suggesting no effect on haematopoietic or immune systems in the rat model (WAG) used. Overall, those studies expressed the advantages of utilising PDLLA as a biomaterial because it possesses advantageous biocompatibility and reduced toxicity [50].

6.4.2 Polyglycolic acid

It has been shown that fast-degrading PGA causes a greater percentage of FBR than slow-degrading PGA implants. An inflammatory response to PGA implants can be observed 12 weeks after implantation due to its degradation rate [8, 57]. Studies on the ability of chondrocytes to be cultured on a fibrous PGA matrix demonstrated that these cells have capacity to adapt and proliferate on this polymer for several weeks after culturing, thereby allowing the formation of cartilage-like tissue with amplified production of collagen [41].

Poly(lactic-*co*-glycolic acid) (PLGA) microspheres are utilised as carriers of sustained release of medication in drug delivery and also utilised as adjuvants [58]. The acute-phase response of PLGA was assessed after the introduction of this copolymer to mouse macrophage-like cells (2D J774). As well as studies carried out in two-dimensional rat peritoneal macrophages and human acute promyelocytic leukemia cells (HL-60), it was observed that PLGA exposure resulted in an increase in levels of important components of the inflammatory cascade, such as IL-1β and TNF-α protein levels [8, 59].

Studies carried out on the use of PLGA root replicas to aid tooth-extraction sockets have shown bone decalcification in the implantation site because of build-up of lactic acid by-products. Similar to PLLA, high-performance liquid chromatography and pH monitoring of media containing PLGA demonstrated pH reduction to be faster than that for PLLA. This finding shows that degradation of PLGA results in a more acidic solution than that for PLLA. An improvement of cell adhesion and growth was observed for PLGA when combined with polyvinyl alcohol (PVA) to form PLGA/PVA scaffolds. *In vivo* analysis using rabbits as a model also demonstrated that PLGA/PVA scaffolds improved bone regeneration in the rabbit skull. When comparing the effect of large and small microspheres of PLGA, it was evident that there was a greater cellular infiltration of smaller PLGA microspheres due to the accessibility of the cells to infiltrate the smaller microspheres. Biocompatibility can be affected by PLGA microsphere size. Analyses *in vivo* using Sprague–Dawley rats revealed larger PLGA microspheres to produce a greater FBR, resulting in cells surrounding the material of the larger microsphere implants. However, macrophages had the capacity to digest smaller particles, resulting in no FBR. No genotoxic potential was observed when human skin fibroblasts were exposed to PLGA. The effects of PLGA on the central nervous system [60], pulmonary system [61], parotid gland [62], the eye [63], excretory system and the heart [64] have been observed after PLGA implantation in Sprague–Dawley rats *in vivo*. Toxicity evaluation of PLGA demonstrated that this biomaterial meets the biocompatibility requirements used in the circulatory system, as well as urologic and otologic applications, and can be utilised as a safe drug-delivery system for ophthalmic applications [8].

6.4.3 Polycaprolactone

Similar to the advantages of PLA, PCL is a widely utilised biodegradable polymer for the production of biodegradable devices in the biomedical industry. Already approved by health authorities such as the FDA, it is a biocompatible polyester [65] with admirable properties such as its ease of processing and an attractive elastic nature. Although it has commendable properties, its degradation results in the formation of acidic products that have the ability to induce inflammation at the site of implantation and with the probability of causing tissue necrosis. It has a high degradation rate and is an expensive polymer. The shape of the implant itself affects the degradation rate of PCL implants. Studies completed by Salgado and co-workers who developed a PCL–sebacic acid (SA) blend showed the biocompatibility and breakdown rate of this injectable gel. Blended with SA (which is an innate dicarboxylic acid), to speed up the PCL degradation rate, the biomaterial had small and few pores with increasing size after the degradation time increased. The degradation observed through biomaterial surface alterations is good in tissue engineering because it permits cell migration *in vivo*, allowing the generation of new tissue at the implantation site. PCL undergoes unstructured stage degradation followed by a loss of mechanical properties after molecular mass reduction [66].

Cytotoxicity analysis of PCL–SA blended gels was carried out utilising MG63 osteoblasts-like cells. These cells were used due to their high proliferation and growth rate and also because of their ease of culture [65]. The MTT assay was used to assess the effect of PCL–SA on cell viability *in vitro*. No decrease in cell viability was observed when MG63 cells were exposed to the PCL–SA blend. When using MG63 cells for direct and indirect assays, the blend was cyto-compatible. Mild toxicity towards the MG63 cells and limited cell death was observed according to ISO 10993-5. MG63 cell morphology, when exposed to PCL–SA, was similar to those exposed to control alone. An indirect assay was also undertaken with dissolution products of the biomaterial demonstrating no significant cell death compared with the cell line incubated with fresh medium and analysed through the same end-point. Indirect and direct contact assays highlighted that PCL–SA gel blends are cyto-compatible. Their high proliferation rate was not affected after exposure to the blend when a cellular layer was observed several days after culturing. Flow cytometric analysis showed the inability of PCL–SA blends to induce apoptosis compared with the control. Scanning electron microscopy images showed the ability of MG63 osteoblast cells to attach and proliferate on PCL–SA gel blends, so they could be used in tissue-engineering applications [13].

Aiming to enhance the biocompatibility and biomechanical properties of PCL, Yu and co-workers completed studies to demonstrate the non-toxic nature of HA–PCL biomaterials *in vitro*. Cytotoxicity studies were carried out utilising bone marrow cells obtained from BALB/c mice. The mitochondrial activity of exposed cells was assessed in biochemistry assays using Alamar blue as an endpoint for cytotoxicity assessment to provide quantitative analysis of the compatibility of this biomaterial *in vitro*. No significant difference in cell viability was observed. Studies have revealed that PCL and HA alone are non-mutagenic and non-toxic [14, 67]. Degradation of PCL *in vivo* results in the generation of two metabolites: ε-hydroxycaproic acid and water [68]. The inclusion of HA to PCL not only enhances the biomechanical properties of the biomaterial but also increases production of nitric oxide, which suggests that HA promotes the functions of endothelial cells. The addition of HA to PCL enhanced the attachment and growth of cells *in vitro* due to alterations to the surface. A rougher surface aided cell attachment and anchorage, allowing improved cell proliferation and differentiation and the generation of new bone. *In vitro*, it was observed that the by-products of HA dissolution neutralised the acidic products produced upon PCL degradation. A compatible environment was achieved when PCL was supplemented with HA for tissue engineering. As well as biochemical assessment, Yu and co-workers characterised cells cultured on HA–PCL scaffolds to demonstrate the ability of the cells to retain their innate morphology after several days [65].

To enhance the degradation rate of PCL, Kweon and co-workers developed a novel PCL macromer and investigated the compatibility of it with a human MG63 osteoblast cell line *in vitro*. The degradation rate was increased through the incorporation of acrylate groups. Cell culture analysis of this PCL network for bone-marrow conductance showed a significant reduction in the viability of osteoblast cells exposed to the PCL macromer compared with the control, but the PCL networks were biocompatible. Hence, this material could be used as a biomedical scaffold for tissue engineering [14].

6.4.4 Polyethylene glycol

Polyethylene glycol (PEG) is one of the most commonly used synthetic polymers for biomedical applications. Through the controlled variation of its MW, water content and monomer concentration, it is possible to fine-tune its materials properties for a range of applications, including structural scaffolds, biodegradable drug-delivery vehicles, and cell carriers. PEG monomer is known widely to be cytotoxic, but PEG as well as its derivatives form biocompatible hydrogels (Figure 6.5) following polymerisation, and have been used in numerous biomedical applications. Furthermore, through the addition of cell adhesive peptides, enzyme sensitive peptides and growth factors, PEG hydrogels have been shown to mimic the function of the extracellular matrix [69–71], allowing for the attachment of multiple cell lines to its surface, including fibroblasts, chondroblasts, osteoblasts and neural cells [72, 73]. These benefits, as well as the ability to add specific factors to improve PEG

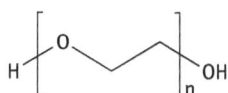

Figure 6.5: Chemical structure of PEG.

degradation times, have led to it becoming a sought-after option in bone, vascular and cartilaginous tissue regeneration [72, 74, 75].

6.4.5 Polyvinyl alcohol

PVA is a synthetic polymer (Figure 6.6) which can be physically crosslinked through the use of multiple freeze–thaw cycles [75] or through chemical crosslinking in combination with certain chemicals, including glutaraldehyde, succinyl chloride or adipoyl chloride [76]. Unlike many synthetic polymers, PVA has inherent cell-adhesion properties. However, these inherent properties were found to be lacking when preparing tissue-engineered scaffolds. PVA has, however, seen improvements to its adhesion. For example, Nuttelman and co-workers improved cell adhesion to fibroblasts through modification of PVA with fibronectin [77]. Schmedlen and co-workers functionalised PVA with cell-adhesive peptides [78]. Furthermore, the non-toxic, non-carcinogenic and rubbery elastic nature of PVA have allowed it to simulate many bodily functions, leading to implementation into areas such as contact lenses, drug delivery and artificial heart linings [79].

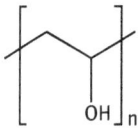

Figure 6.6: Chemical structure of PVA.

6.4.6 Polyvinylpyrrolidone

Polyvinylpyrrolidone (PVP) is a water-soluble synthetic polymer that can be prepared from N-vinylpyrrolidone. PVP is one of the most common synthetic polymers along with the aforementioned PEG and PVA. Following PVP creation, it was initially utilised as a plasma expander and has since seen its applications spread into many fields, such as the medical and food industries. Due to its chemistry (Figure 6.7), PVP has a very high affinity for water, which leads to its rapid breakdown if in

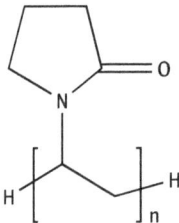

Figure 6.7: Chemical structure of PVP.

contact with water [80]. Thanks to this rapid degradation the biomedical applications of PVP are numerous, including the addition to other polymers to fine tune their degradation rates for tissue-engineering scaffolds [81] and separately to act as a model for rapid drug delivery [82].

6.5 Summary

Many factors influence the potential FBR generated after biomaterial implantation. Various influential factors can be controlled during the production of these synthetic bioresorbable polymers for tissue engineering. Using resorbable materials for GTR prevents the need for implant removal because these biomaterials have the capacity to be broken down and eliminated from the body. Studies have demonstrated that the most commonly utilised bioresorbable polymers for tissue engineering, such as PLA, PCL and PGA, have reduced toxicity, are readily biodegradable, and are biocompatible, which makes then admirable biomaterials for implantation in the human body. The potential use of bioresorbable polymers for the production of biological devices to aid bone-tissue engineering and skin engineering is an exceedingly positive development.

References

1. J.M. Anderson, *Annual Review of Materials Research*, 2001, **31**, 81.
2. A. Asti and L. Gioglio, *The International Journal of Artificial Organs*, 2014, **37**, 187.
3. J.M. Morais, F. Papadimitrakopoulos and D.J. Burgess, *American Association of Pharmaceutical Scientists*, 2010, **12**, 188.
4. Y. Ramot, A. Nyska, E. Markovitz, A. Dekel, G. Klaiman, M.H. Zada, A.J. Domb and R.R. Maronpot, *Toxicologic Pathology*, 2015, **43**, 1127.
5. A. Nyska, Y.S. Schiffenbauer, C.T. Brami, R.R. Maronpot and Y. Ramot, *Polymers for Advanced Technologies*, 2014, **25**, 461.
6. Y. Onuki, U. Bhardwaj, F. Papadimitrakopoulos and D.J. Burgess, *Journal of Diabetes Science and Technology*, 2008, **2**, 1003.
7. E. Fournier, C. Passirani, C.N. Montero-Menei and J.P. Benoit, *Biomaterials*, 2003, **24**, 3311.
8. Y. Ramot, M.H. Zada, A.J. Domb and A. Nyska, *Advanced Drug Delivery Reviews*, 2016, **15**, 153.
9. A. Boccaccini, *Composites Science and Technology*, 2003, **63**, 2417.
10. S. Chung, N.P. Ingle, G.A. Montero, S.H. Kim and M.W. King, *Acta biomaterialia*, 2010, **6**, 1958.
11. D. Brie, P. Penson, M-C. Serban, P.P. Toth, C. Simonton, P.W. Serruys and M. Banach, *International Journal of Cardiology*, 2016, **215**, 47.
12. C. Liu, K.W. Chan, J. Shen, H.M. Wong, K.W. Kwok Yeung and S.C. Tjong, *RSC Advances*, 2015, **5**, 72288.
13. C.L. Salgado, E.M.S. Sanchez, C.A.C. Zavaglia and P.L. Granja, *Journal of Biomedical Materials Research: Part A*, 2012, **100**, 243.
14. H. Kweon, M. Kyong, I. Kyu, T. Hee, H. Chul, H. Lee, J. Oh, T. Akaike and C. Cho, *Biomaterials*, 2003, **24**, 801.

15. H. Baumann and J. Gauldie, *Immunology Today*, 1994, **15**,74.
16. M. Generali, P.E. Dijkman and S.P. Hoerstrup, *European Medical Journal*, 2014, **1**, 91.
17. A. Terriza, J.I. Vilches-pérez, E. De Orden, F. Yubero, J.L. Gonzalez-Caballero, A.R. González-Elipe, J. Vilches and M. Salido, *BioMed Research International*, 2014, **2014**, 1.
18. K.Y. Lee and D.J. Mooney, *Progress in Polymer Science*, 2013, **37**, 106.
19. E. Ivanova, K. Bazaka and R. Crawford in *Natural Polymer Biomaterials: Advanced Applications*, Woodhead Publishing, Cambridge, UK, 2014, p.32.
20. M. George and T.E. Abraham, *Journal of Controlled Release*, 2006, **114**, 1.
21. J. Li, J. He and Y. Huang, *International Journal of Biological Macromolecules*, 2017, **94**, 466.
22. D. Silva, L.F. V Pinto, D. Bozukova, L.F. Santos, A. Paula and B. Saramago, *Colloids and Surfaces B: Biointerfaces*, 2016,**147**, 81.
23. R. Raguvaran, B.K. Manuja, M. Chopra, R. Thakur, T. Anand, A. Kalia and A. Manuja, *International Journal of Biological Macromolecules*, 2017, **96**, 185.
24. G. Orive, S.K. Tam, J.L. Pedraz and J.P. Hallé, *Biomaterials*, 2006, **27**, 3691.
25. G. Orive, S. Ponce, R.M. Hernández, A.R. Gascón, M. Igartua and J.L. Pedraz, *Biomaterials*, 2002, **23**, 3825.
26. S.K. Tam, J. Dusseault, S. Bilodeau, G. Langlois, J.P. Hallé and L. Yahia, *Journal of Biomedical Materials Research: Part A*, 2011, **98A**, 40.
27. P.J. Vandevord, H.W.T. Matthew, S.P. Desilva, L. Mayton, B. Wu and P.H. Wooley, *Journal of Biomedical Materials Research*, 2002, **59**, 585.
28. F. Croisier and C. Jérôme, *European Polymer Journal*, 2013, **49**, 780.
29. J. Venkatesan, J-Y. Lee, D.S. Kang, S. Anil, S-K. Kim, M.S. Shim and D.G. Kim, *International Journal of Biological Macromolecules*, 2017, **98**, 515.
30. D. Ozdil, I. Wimpenny, H.M. Aydin and Y. Yang in *Science and Principles of Biodegradable and Bioresorbable Medical Polymers*, 1st Ed, Woodhead Publishing, Cambridge, UK, 2017.
31. S. Rodrigues, M. Dionísio, C.R. López and A. Grenha, *Journal of Functional Biomaterials*, 2012, **3**, 615.
32. L. Zhao, H.J. Gwon, Y.M. Lim, Y.C. Nho and S.Y. Kim, *Radiation Physics and Chemistry*, 2015, **106**, 404.
33. Y. Zhang, P. Heher, J. Hilborn, H. Redl and D.A. Ossipov, *Acta Biomaterialia*, 2016, **38**, 23.
34. A. Sionkowska, B. Kaczmarek, K. Lewandowska, S. Grabska, M. Pokrywczyńska, T. Kloskowski and T. Drewa, *International Journal of Biological Macromolecules*, 2016, **89**, 442.
35. M. Hemshekhar, R.M. Thushara, S. Chandranayaka, L.S. Sherman, K. Kemparaju and K.S. Girish, *International Journal of Biological Macromolecules*, 2016, **86**, 917.
36. S. Wang, S. Guan, J. Wang, H. Liu, T. Liu, X. Ma and Z. Cui, *Journal of Bioscience and Bioengineering*, 2016, **123**, 116.
37. P. Bulpitt and D. Aeschlimann, *Journal of Biomedical Materials Research*, 1999, **47**, 152.
38. M. Khanmohammadi, S. Sakai and M. Taya, *International Journal of Biological Macromolecules*, 2017, **97**, 308.
39. Y. Xu, S. Asghar, L. Yang, H. Li, Z. Wang, Q. Ping and Y. Xiao, *Carbohydrate Polymers*, 2017, **157**, 419.
40. D. Moorkoth and K.M. Nampoothiri, *Applied Biochemistry and Biotechnology*, 2014, **174**, 2181.
41. A.R. Santos, Jr., in *Tissue Engineering*, Ed., D. Eberli, InTech, Rijeka, Croatia, 2010, p.225.
42. Y. Chen, L.M. Geever, J.A. Killion, J.G. Lyons, C.L. Higginbotham and D.M. Devine, *Polymer–Plastics Technology and Engineering*, 2016, **55**, 1057.
43. K. Rezwan, Q.Z. Chen, J.J. Blaker and A.R. Boccaccini, *Biomaterials*, 2006, **27**, 3413.
44. L. Xiao, B. Wang, G. Yang and M. Gauthier in *Biomedical Science, Engineering and Technology*, Ed., D.N. Ghista, InTech, Rijeka, Croatia, 2012, p.249.

45. I. V. Maiborodin, I. V. Kuznetsova, E.A. Beregovoi, A.I. Shevela, M.I. Barannik, V.I. Maiborodina and A.A. Manaev, *Bulletin of Experimental Biology and Medicine*, 2014, **156**, 874.
46. A. Stojadinovic, J.W. Carlson, G.S. Schultz, T.A. Davis and E.A. Elster, *Gynecologic Oncology*, 2008, **111**, S70.
47. K. Nair, *Pushpagiri Medical Journal*, 2011, **3**, 19.
48. N. Thomas, G. Sanil, R. Gopimohan, J. Prabhakaran, G. Thomas and A. Panda, *Journal of Indian Society of Periodontology*, 2012, **16**, 498.
49. M. Mohiti-Asli, S. Saha, S.V. Murphy, H. Gracz, B. Pourdeyhimi, A. Atala and E.G. Loboa, *Journal of Biomedical Materials Research, Part B: Applied Biomaterials*, 2015, **105**, 327.
50. R. Li, Z. Liu, H. Liu, L. Chen, J. Liu and Y. Pan, *American Journal of Translational Research*, 2015, **7**, 1357.
51. G. Sui, X. Yang, F. Mei, X. Hu, G. Chen, X. Deng and S. Ryu, *Biomedical Materials Research*, 2007, **82A**, 445.
52. R.E. Neuendorf, E. Saiz, A.P. Tomsia and R.O. Ritchie, *Acta Biomaterialia*, 2008, **4**, 1288.
53. L. Shen, H. Yang, J. Ying, F. Qiao and M. Peng, *Journal of Materials Science: Materials in Medicine*, 2009, **20**, 2259.
54. W. Zhu, D. Guo, Y. Chen, W. Xiu, D. Wang, J. Huang, J. Huang, W. Lu, L. Peng, K. Chen and Y. Zeng, *Artificial Cells, Nanomedicine, and Biotechnology*, 2016, **44**, 1122.
55. G. Gonzalez-Lomas, R.T. Cassilly, F. Remotti and W.N. Levine, *Clinical Orthopaedics and Related Research*, 2011, **469**, 1082.
56. G.S. Athwal, S.M. Shridharani and S.W. O'Driscoll, *The Journal of Bone and Joint Surgery: American Volume*, 2006, **88**, 1840.
57. O. Bostman, E. Hirvensalo and J. Makinen, *The Journal of Bone and Joint Surgery*, 1990, **72B**, 4, 592.
58. W. Jiang, R.K. Gupta, M.C. Deshpande and S.P. Schwendeman, *Advanced Drug Delivery Reviews*, 2005, **57**, 391.
59. R. Nicolete, D.F. dos Santos and L.H. Faccioli, *International Immunopharmacology*, 2011, **11**, 1557.
60. D.Y. Lewitus, K.L. Smith, W. Shain, D. Bolikal and J. Kohn, *Biomaterials*, 2011, **32**, 5543.
61. L.A. Dailey, N. Jekel, L. Fink, T. Gessler, T. Schmehl, M. Wittmar, T. Kissel and W. Seeger, *Toxicology and Applied Pharmacology*, 2006, **215**, 100.
62. M. Cantín, P. Miranda, I.S. Galdames, D. Zavando and P. Arenas, *International Journal of Clinical and Experimental Pathology*, 2013, **6**, 2412.
63. X. Rong, S. Yang, H. Miao, T. Guo, Z. Wang, W. Shi, X. Mo, W. Yuan and T. Jin, *Investigative Ophthalmology & Visual Science*, 2012, **53**, 6025.
64. M. Szymonowics, Z. Rybak, W. Witkiewicz, C. Pezowics and J. Filipiak, *Acta of Bioengineering and Biomechanics*, 2014, **16**, 131.
65. H. Yu, P.H. Wooley and S-Y. Yang, *Journal of Orthopaedic Surgery and Research*, 2009, **4**, 5.
66. D.R. Chen, J.Z. Bei and S.G. Wang, *Polymer Degradation and Stability*, 2000, **67**, 455.
67. H. Huatan, J.H. Collett, D. Attwood and C. Booth, *Biomaterials*, 1995, **16**, 1297.
68. L.G. Griffith, *Acta Materialia*, 2000, **48**, 263.
69. J. Zhu, *Biomaterials*, 2010, **31**, 4639.
70. M.C. Cushing and K.S. Anseth, *Science*, 2007, **316**, 1133.
71. C.R. Nuttelman, M.A. Rice, A.E. Rydholm, C.N. Salinas, D.N. Shah and K.S. Anseth, *Progress in Polymer Science (Oxford)*, 2008, **33**, 167.
72. Z.A.A. Hamid and K.W. Lim, *Procedia Chemistry*, 2016, **19**, 410.
73. T. Fukumori and T. Nakaoki, *Open Journal of Organic Polymer Materials*, 2013, **3**, 110.
74. J.A. Killion, S. Kehoe, L.M. Geever, D.M. Devine, E. Sheehan, D. Boyd and C.L. Higginbotham, *Materials Science and Engineering C: Materials for Biological Aapplications*, 2013, **33**, 4203.
75. P. Kim, A. Yuan, K-H. Nam, A. Jiao and D-H. Kim, *Biofabrication*, 2014, **6**, 24112.

76. J.L. Drury and D.J. Mooney, *Biomaterials*, 2003, **24**, 4337.
77. C.R. Nuttelman, D.J. Mortisen, S.M. Henry and K.S. Anseth, *Journal of Biomedical Materials Research*, 2001, **57**, 217.
78. R.H. Schmedlen, K.S. Masters and J.L. West, *Biomaterials*, 2002, **23**, 4325.
79. C.M. Hassan and N.A. Peppas, *Advances in Polymer Science*, 2000, **153**, 37.
80. B. Jirgensons, *Journal of Polymer Science*, 1952, **8**, 519.
81. G.M. Kim, K.H.T. Le, S.M. Giannitelli, Y.J. Lee, A. Rainer and M. Trombetta, *Journal of Materials Science: Materials in Medicine*, 2013, **24**, 1425.
82. D-G. Yu, X-X. Shen, C. Branford-White, K. White, L-M. Zhu and S.W.A. Bligh, *Nanotechnology*, 2009, **20**, 55104.

Abbreviations

3D	Three-dimensional
4-ASA	4-Aminosalicylic acid
5-ASA	5-Aminosalicylic acid
5-FU	5-Fluorouracil
AED	Anti-epileptic drugs
AgNP	Silver nanoparticle(s)
ALP	Alkaline phosphatase
AML	Acute myeloid leukaemia
API	Active pharmaceutical ingredients
AuNP	Gold nanoparticle(s)
BBB	Blood–brain barrier
BMS	Bare metallic stents
BVS	Biodegradable vascular scaffold
CABG	Coronary artery bypass graft
CaP	Calcium phosphates
CD	Cyclodextrins
CHD	Coronary heart disease
CMC	Carboxymethyl cellulose
CO_2	Carbon dioxide
CS	Chitosan
DBM	Demineralised bone matrix
DES	Drug-eluting stents
Dex	Dexamethasone
DNA	Deoxyribonucleic acid
EC	Ethyl cellulose
ECM	Extracellular material
EGFR	Epithelial growth factor receptor
FBGC	Foreign body giant cells
FBR	Foreign-body reaction
FDA	Food and Drug Administration (US)
FDM	Fused deposition modelling
GDNF	Glial cell line-derived neurotrophic factors
GIT	Gastrointestinal tract
GTR	Guided tissue regeneration
HA	Hydroxyapatite
HCAEC	Human coronary artery endothelial cells
HCASMC	Human coronary artery smooth muscle cells
HDF	Human dermal fibroblasts
HEK	Human embryonic kidney
HGC	Hexanoyl glycol chitosan
HIV	Human immunodeficiency virus
HME	Hot-melt extrusion
hMSCs	Human mesenchymal stem cells
HNT	Halloysite nanotube
IC_{50}	Half-maximal inhibitory concentration
IL	Interleukin
ISO	International Organization for Standardization

https://doi.org/10.1515/9783110640571-007

LbL	Layer-by-layer
LCST	Lower critical solution temperature
MA	Maleic anhydride
MNP	Multifunctional nanoparticle
MRI	Magnetic resonance imaging
MW	Molecular weight
MWD	Molecular weight distribution
NECL1	Nectin-like molecule 1
nHA	Nano-hydroxyapatite
NP	Nanoparticle(s)
NVCL	N-vinylcaprolactam
PAA	Polyacrylic acid
PAH	Polyallylamine hydrochloride
PBCA	Polybutylcyanoacrylate
PBT	Polybutylene terephthalate
PC	Polycarbonate
PCI	Percutaneous coronary intervention
PCL	Polycaprolactone
PCLA	Poly(ε-caprolactone lactide)
PDLA	Poly(D-lactic acid)
PDLLA	Poly(D,L-lactic acid)
pDNA	Plasmid deoxyribonucleic acid
PDO	Polydioxanone
PE	Polyethylene
PECE	Poly(ε-caprolactone)-polyethylene glycol-poly (ε-caprolactone)
PEG	Polyethylene glycol
PEO	Polyethylene oxide
PET	Polyethylene terephthalate
PGA	Polyglycolic acid(s)
P-gp	P-glycoprotein
PHA	Polyhydroxyalkanoates
PHB	Polyhydroxybutyrate
PLA	Polylactic acid
PLGA	Poly(lactic-*co*-glycolic acid)
PLLA	Poly(L-lactic acid)
PNIPAAm	Poly(N-isopropylacrylamide)
PNS	Peripheral nerve system
PNVCL	Poly(N-vinylcaprolactam)
PP	Polypropylene
PPF	Polypropylene fumarate
PPO	Polypropylene oxide
Pq	Pyrroloquinoline quinone
Psi	Porous silicone
PtNP	Platinum nanoparticle(s)
PTX	Paclitaxel
PU	Polyurethanes
PVA	Polyvinyl alcohol
PVP	Polyvinylpyrrolidone
RAFT	Reversible addition-fragmentation transfer

RBC	Red blood cells
RG2	Rat glioma 2
RGD	Arg-Asp-D-Phe-Cys
ROS	Reactive oxygen species
RSV	Resveratrol
SA	Sebacic acid
siRNA	Small-interfering ribonucleic acid
TDP	Thermal drawing processing
T_g	Glass transition temperature
TGA	Thermogravimetric analysis
THF	Thyrotropin-releasing factor
T_m	Melting temperature
TMC	Trimethylene carbonate
TNF	Tumour necrosis factor
TPLE	Paclitaxel-loaded lipid nanoemulsion
β-TCP	Beta-tricalcium phosphate
ε-CL	ε-Caprolactone

Index

https://doi.org/10.1515/9783110640571-008

www.ingramcontent.com/pod-product-compliance
Lightning Source LLC
Chambersburg PA
CBHW081540220326
41598CB00036B/6496